GPS Praxis Book Series

GPS Praxis Book
GARMIN GPSMAP 64 Series

All rights reserved. No part of this publication may be reproduced in any way.

The information and descriptions in this manual are included solely for informational purposes and are subject to change.

Most of the product descriptions of hardware and software, as well as company names and logos mentioned in this work, are generally also registered trademarks and should be treated as such. They are mentioned for identification purposes only and are the exclusive property of their holders.

The compilation of texts and illustrations was undertaken with the greatest of care. Nevertheless, errors cannot be completely ruled out. The publisher and the author accept no liability, or guarantee, and also take no responsibility for consequential damages due to errors, or inaccuracies in this manual.

Production & Publishing:
BoD - Books on Demand, Norderstedt

Author und Graphics: Janet Bader

English translation: Dustin Baer, Sept.2015

GPS Praxis Book Garmin GPSMAP64 Series
Edition 2 – October 2016
© 2016 Red Bike

ISBN 978-3-7386-1494-7

Table of Contents

Chapter-Page

FOREWORD 7

BASIC EQUIPMENT 8

CHAPTER 1 - GENERAL 1–10

    OPERATION OF THE GPSMAP 64 SERIES 1–10
    DIFFERENT TYPES OF MAPS 1–12
        Road maps *1–12*
        Topographic maps *1–13*
        BirdsEye *1–14*
        Nautical maps *1–18*
    GARMIN SOFTWARE: MAPSOURCE, BASECAMP, CONNECT 1–20
    ROUTES AND TRACKS 1–22
    TRACK POINTS, WAYPOINTS, VIA POINTS AND POIS 1–24
    GEOCACHING 1–26
    PHOTO TAGGING 1–33
    COORDINATE SYSTEM 1–34
        North reference / Declination *1–37*
        GPS + GLONASS *1–38*
        WAAS und EGNOS *1–38*
    UPDATES 1–39

CHAPTER 2 - THE DEVICE 2–41

    INSERTING BATTERIES AND TURNING ON THE DEVICE 2–41
    THE KEYS AND THEIR MEANING 2–42
        Overview of keys and key shortcuts *2–47*
        Working in the map view *2–49*
        Working in the Elevation Plot view *2–50*
    BASIC SETTINGS 2–52
    OTHER MEANINGFUL SETTINGS 2–58
        With different types of uses *2–66*
    BLUETOOTH – UTILIZATION 2–68
        Receiving messages *2–68*
        Live-Tracking *2–69*
        Transmitting GPS Elements *2–72*
        Pairing to a VIRB™ Action Camera *2–76*

CHAPTER 3 - NAVIGATION **3–79**
  ROUTE NAVIGATION 3–80
    Tour start - Route *3–82*
    Changing the active route *3–84*
    Prepare a route in the device *3–84*
  TRACK NAVIGATION 3–86
    Send data from a PC to the GPS64, without GPS software *3–88*
    Transfer via the ANT+ interface *3–91*
    Switch on Track visibility *3–92*
    Tour start - Track *3–94*
    Track with turn-by-turn instructions *3–95*
  SIGHT 'N GO 3–97
  TRACBACK 3–99
  TOUR START/TOUR END - STEPS IN THE DEVICE 3–100
    Save the Track record *3–100*

CHAPTER 4 – WORKING ON A PC **4–105**
  GARMIN FILE FORMATS: GPX, GDB, FIT, TCX, CRS 4–105
  CREATE A BACKUP COPY OF THE GPS DEVICE STORAGE 4–106
  DEVICE STORAGE: SYSTEM AND FOLDER STRUCTURE 4–106
  MICROSD CARD SET UP 4–108
  MAP INSTALLATION 4–110
    Preprogrammed map data - microSD card *4–110*
    Map data - DVD installation on a PC *4–111*
    Sending maps from a PC to the GPS device *4–113*
  TOURS FROM THE "NET" 4–117
  PLANNING AND DRAWING TOURS YOURSELF 4–120
    Drawing in BaseCamp *4–122*
    The search function in BaseCamp *4–130*
    Other planning programs *4–135*
    Elevation values: barometric, via GPS, or from the map *4–137*
  CREATING WAYPOINTS IN BASECAMP 4–138
    Create waypoints via coordinates *4–139*
    Create your own waypoint icon *4–141*
  GEOREFERENCE PHOTOS 4–142
  EVALUATE TRACK RECORDING ON THE PC 4–145
    Open a recording in BaseCamp *4–145*
    Open a recording in Garmin Connect *4–150*

INDEX **154**

Foreword

Welcome to the circle of outdoor GPS device users, who have opted for an all-purpose device that is designed for use in harsh environments in the field and is easy to operate, even with gloves. In short: the GPSMAP 64 model series outdoor navigation device, in the proven and now already iconic body of the first edition.

As a Garmin dealer in the category of sports and outdoors and as a cycling-sport infected bike retailer, we link many routes with the use of GPS technology ourselves and often have your exact type of device on tour. This gives us the ability to give you step-by-step descriptions of how it should work in practice.

The basic operations described here are fully applicable to all GPSMAP 64 devices with the model suffix "s," and also applicable to devices without the "s" extension, with some exceptions. Since the latter doesn't have a barometric altimeter, 3-axis compass, no wireless connection, there is no way to pair heart rate and cadence sensors, or smart phones, thus, explanations on these topics logically don't apply. However, all devices have completely the same navigation abilities and data structure. The abbreviated terms "GPS64" or " '64" are used often in the book, when discussing the entire series of devices.

Gradually, you will surely discover other ways you can use and adapt your device to better suit your needs and even long after purchase, be surprised what can be done with your "'64". To discuss such details here is beyond the scope of this book and would make the reading for GPS beginners absolute agony.

Basic equipment

Getting started with:

- **GPS device** – GPSMAP 64...

- **Maps** for the GPSMAP 64. Best are routable, so that it can automatically calculate the route to the destination.

For working on a PC:

- **GPS map software** "BaseCamp," to create and edit routes and for data transfer between the GPS device and a PC/Mac. For installation on a computer.
  Download: www.garmin.com > Support > Software

- **User account** for "Garmin Connect" – the worldwide training portal for sports enthusiasts – for detailed evaluation of fitness data (such as heart rate and cadence), as well as usage of the live tracking function. Online: http://connect.garmin.com (choose your language)

For the most current software version:

- **"Garmin Express,"** in order to search for the latest updates for the GPS64. For installation on a computer. Download: www.garmin.com > Support > Software

To unlock a Garmin map DVD:

- **"Garmin user account"** – in order to unlock Garmin maps for the PC and the device, as well as to use other services.
  Online: www.garmin.com > Sign In

Chapter 1 - General

Operation of the GPSMAP 64 Series

The models in the GPSMAP 64 series are now clearly seen as **the** cult outdoor GPS device. With the first edition of the GPSMAP 60, a usable outdoor GPS device came onto the market for the first time. With the introduction of the highly sensitive Sirf III receiver, which was first built in the GPSMAP 60CSx, the visually distinctive unit arrived with great fame and recognition. Thus, it had been burned into minds and the useful design established. That is likely the reason why the exact identical device is now experiencing its 3rd edition with the glove-enabled keys.

All models of the '64 series always offer plenty of local knowledge while hiking, climbing, bicycle riding, motorcycling, or car driving, paragliding, hang gliding and treasure hunting (geocaching). They also act as an unassuming guide at a completely unknown holiday destination, or as a navigator at sea. Of course, it's assumed that the device was "fed" with the corresponding data (maps), before heading out. For this, there is now a wide selection of readily available, routable topographical maps, road maps and nautical charts that can be used with a simple "turn on and go."

The Garmin outdoor models have a robust, impact-resistant plastic housing and are waterproof to IPX7 standards (30-minute submersion in 1 meter/3 feet deep water, no salt water), but they are not buoyant. The '64 models can withstand temperatures between -20 and +70°C (-4 and +158°F), which is below/above the recommended temperature range of many batteries. All models of the new '64 series are equipped with high-sensitivity receiver chips. Dense forest and narrow canyons can hardly bring these devices out of balance.

All '64 models with an "s" in the model name, feature the following interfaces:

- "ANT+" Technology, which gives you the ability to wirelessly transmit data to and from another ANT+ enabled device. This wireless data transmission also allows you to take advantage of the optional heart rate belt, the cadence and speed sensor ("GSC 10") for use with a bike, the Geocaching "Chirp" transmitter, the temperature sensor ("tempe") and the Garmin GPS Action Camera ("VIRB").
- the serial USB interface, which is used as the main data exchange for PC/Mac and
- the Bluetooth interface, through which you can establish a connection to a smartphone, which can then show messages, appointments and calls on the GPS64's display, manage all your GPS movement and send data to other observers.

The storage of all '64 models can be expanded by up to 32 GB with a microSD card. Of this, only 4 GB of map data, or 2,025 map tiles can be read by your device. Additional space can be used for all photo files and other GPS-objects. Apart from that, the devices themselves offer ample internal storage to, for example, record its own data.

The device comes with a worldwide, shaded-relief basemap built in. The map only displays interstates, highways, cities (as dots) and large bodies of water. The '64 models, with the small "t" at the end of the model name, come with a recreational TOPO map of your country, or continent (Europe, U.S.A., Canada, Australia & New Zealand, etc.), at a scale of 1:100,000 with contour lines at intervals of 25 meters (80 feet). However, no automatic routing calculation is available on either map.

With the GPS64, you can be guided along a route of a prepared tour, or can automatically navigate to a destination selected in the device. For the latter variant, routable maps are required. The programmed route must be compatible with the device and for the relevant activity type, e.g. pedestrian walking, cycling, driving, etc.

These types of maps are already available in various equipment packages, e.g. GPSMAP 64s with TOPO TransAlpin microSD/SD.

In short, only maps from Garmin, or maps that are converted for use with Garmin, are possible. There is a distinction between three

Different types of maps

There are nautical maps, topographical maps and road maps. The latter include all paved roads, sometimes even frequently travelled gravel roads, as well as countless, useful information about certain addresses, a.k.a. Points of Interest (POI), such as attractions, sports and leisure facilities, from accommodations to hospitals and their emergency numbers. These

Road maps can be purchased as a DVD for installation on a PC, or as a pre-programmed microSD card. Installing the map files from the DVD onto a PC has the advantage that you can get a very good overview on the PC screen before the trip and then comfortably plan ahead for all sorts of possibilities, without having to connect the GPS device to the PC. If you have multiple map DVDs, you can upload the map files from the various maps onto the GPS64's microSD card. The map DVDs are normally licensed for use with only 1 computer and 1 GPS unit and they must be activated online.

On the other hand, the preprogrammed microSD cards (as contained in some device packages), are ready for immediate use and can be read by several GPS devices. These regionally preprogrammed data cards come available with the same maps as the DVDs, as well as scaled-down versions with a smaller coverage area for a lower price. Use of these map files with a PC is only possible with a connected GPS64 and with the "BaseCamp" map software from Garmin installed. BaseCamp can then easily access the map files from the microSD card, while it is the device. However, it is also possible to use the map files in BaseCamp by inserting the microSD card into the adapter that came with it and then the adapter into an available card reader slot on your home computer.

Garmin road maps are routable, which means it's possible to automatically calculate a route to a specific destination.

Topographic maps - also known as recreational, or hiking maps - include roads, paths and trails, bodies of water, vegetation, land forms, peaks and numerous POIs such as hotels, restaurants, mountain huts, etc. In the meantime, it is now possible for the GPS64 to calculate the way to a selected destination with these maps as well. All Garmin TOPO maps, which have been released since 2010, have the capability for automatic navigation and can use the routing function. Beginning with the "... 2012 Pro" versions, routing can even be adjusted to the particular activity, e.g. hiking, climbing, mountain biking, tour biking, etc. These cards are advertised with the so-called "ActiveRouting." In the latest release of the Garmin "TOPO Germany V6 PRO" map, it is now even possible to search for a street and house number.

➜ However, using a topographical map in a car is to be treated with caution. Despite the use of activity type "Automobile Driving," it may happen that the route calculation always prefers the shorter route, without complying with the local traffic regulations. With these maps, it can happen that you are directed to drive the opposite direction in roundabouts, and with highway on-ramps, the map could quickly instruct you to use the much closer off-ramp. ←

These maps are also available as a DVD for the licensed use on a PC and as an immediate-use microSD card for the GPS device. It can vary that some regions are only available as a set, while others are available either as a DVD, or microSD/SD card. Due to their detailed representation, the coverage area of these cards is much smaller, as compared to road maps.

Topographic maps, in electronic form, can also exist page-by-page, such as the Garmin Alpine Club maps. All currently available map pages of the Austrian Alpine Club are included and have been optimized and digitized for the Garmin GPS device. With their very high level of detail, they serve especially well as a planning basis for alpinists. These map files are all stored on a microSD card.

By contrast, Garmin provides a full-coverage map image with the material of the KOMPASS-Karten GmbH (map company based in Germany) on microSD card, such as the Garmin "Wander-Atlas Tirol" (Tirol hiking map). As one already knows from the KOMPASS paper maps, they provide a very simple and familiar map image that colorfully highlights hiking, cycling and mountain bike paths. Likewise, some suggested routes are included. The map is routable and includes elevation information with which you can, in advance, calculate the altitude that must be conquered. The price of €40 ($45 US) will probably appear more acceptable to some, as compared to a map of the entire Alpine region of that has to be purchased for €180 ($195 US).

BirdsEye (map images)

The appearance of Garmin maps is made possible by the use of vector maps, which require much less space and ensures a clear and sharp display at all zoom levels. This may take some getting used to and some GPS users may prefer to have an image of a typical hiking map made out of conventional paper in front of them (apart from the battered folding). For this reason, you have the opportunity to use the BirdsEye Select service, at the price of €20 ($25 US), to select up to 2,400 km² (927 mi²) of map sections of a country via the BirdsEye Wizard in the "BaseCamp" software on your PC and to upload it to the GPS64. In conjunction with a routable vector map in the device, you are then able to use the automatic route calculation and the familiar look of a raster map, at the same time.

Those who prefer to have a bird's-eye view of the surroundings of their planned activity will be well served with BirdsEye Satellite Imagery . For a yearly subscription price of €25 ($30 US) you can download an unlimited number of high-resolution satellite images and upload them to the '64, depending on the amount of disk space on it, or its memory card. These will "melt" with the vector maps that are built into the unit and portray a realistic view from a bird's-eye perspective of roads, buildings and terrain. This gives you the ability to locate a suitable parking spot at the starting point of a tour, yet still use the automatic calculation function of the vector map.

➔ Owners of the "s" models can use this service free of charge for the first year. ⬅

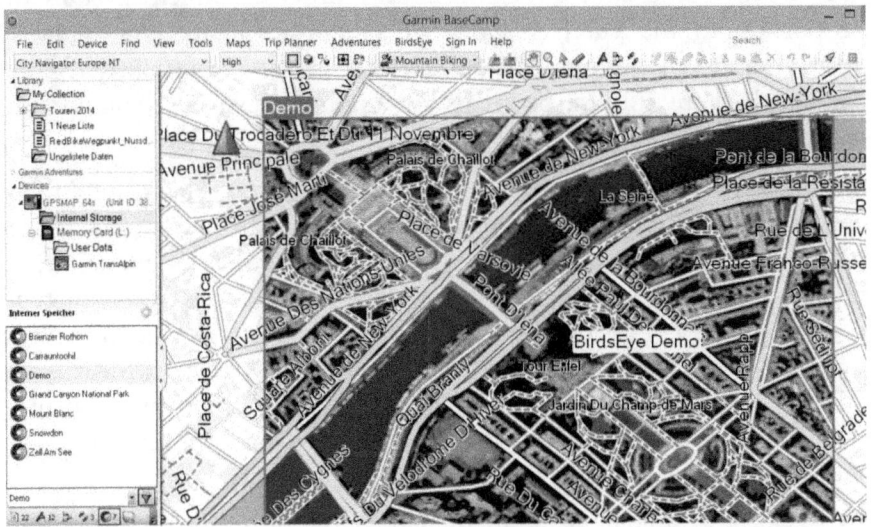

Figure 1-1 The BirdsEye Satellite image and the routable road map melt together. Shown here: BaseCamp Software on the PC

You activate this service via the wizard in the BaseCamp software (menu bar: BirdsEye -> Download BirdsEye Imagery...). If BirdsEye Imagery was not included with the purchase of your GPS model, you will be forwarded to a product description where you will also find information on currently available coverage, if this. The subscription is bound to a single device, like most Garmin maps.

Now back to the map types, because with the <u>Custom Maps</u> Function, there is another way non-Garmin map material can be used in the GPS64. Maps that are self-scanned, or those obtained from other various sources, royalty-free, can be georeferenced using the free

1–15

version of Google Earth (align and adjust the coordinates) and then converted to Garmin-compatible maps. For instructions, we have placed a PDF for free download (in German): http://www.red-bike.de/Zugabe/RB_PraxBuch_CustomMaps.pdf

This feature is especially useful if you want to use your own maps of activities in the GPS device. For widespread use, however, this is a far too cumbersome and time-consuming process, which is out of proportion to the purchase price of a high-resolution and routable vector map from Garmin. If necessary, it might be for a holiday region, for which there still are no maps for Garmin devices...a solution, at least.

For those looking for missing maps, OpenStreetMaps (OSM) is certainly a recommended address on the web. This site offers a free world map in the spirit of Wikipedia. The spatial data originates from GPS logs and measurements and is managed by industrious amateur cartographers worldwide, who have laid their own network of routes over our globe. The maps from OSM can therefore be used for free on a PC and the GPS device, but also expanded through private participation. One can sometimes stumble upon maps from popular regions, or metropolitan areas, which are made of superb quality.

Especially for holiday areas, for which there are no digital maps, one can often find a map deposited for their Garmin device. Maps for a variety of devices, regions and types of use (MTB, cycling, hiking, etc.), are sometimes available on various other portals, which are linked from the OSM site and reached by clicking on the appropriate map selection. Therefore, the path to download a corresponding map may be different every time and changes constantly with the continuously increasing number of people interested in this activity.

For OSM-newbies, as a first step, it is highly recommended to type the direct link into your browser: "http://wiki.openstreetmap.org/wiki/Main_Page". If not already done, you can change the page appearing in the top bar to your own language. On this page you'll find the directory, which clearly describes the first steps and possibilities in

OSM. So, if you want to use maps in a Garmin GPS, click the "Using OpenStreetMap" link and scroll down the page to "Maps on your GPS device." There, you can choose the link to "OSM Map On Garmin."

A map download for an area with good coverage can quickly become 1GB in size. Herein is usually an installation file (.exe), which automatically integrates the map into the Garmin GPS software on the PC. On the other hand, an OSM map might also only be a pure "IMG" map file, which you have to copy into the "Garmin" folder in the GPS64 device storage, or the microSD card via the desktop explorer.

OSM maps are pixel maps with large storage requirements and a stored "vector backbone," ultimately with which a Garmin device can also calculate/create a route to a destination, based on the registered routes within the map.

The disadvantage to Garmin maps: Elevation data is not always integrated into OSM maps. Thus, a representation of an elevation profile is often not possible during the planning of a route on a PC, or the start of a route on the GPS device. The visibility of the OSM map on the device display is a bit more arduous, especially in metropolitan areas. In order to see the guided route, or waypoints, you need to look very closely through all the details, which you are not able to blend out. Similarly, it happens every now and then, that one, or another functions are overridden in the BaseCamp software.

Of course, you can also use

Nautical maps, such as Garmin's "BlueChart g2," in the GPS64 by means of a preprogrammed microSD card. These maps include, for example, realistic navigation features, nautical navigation aids such as currents, IALA map symbols, marine POIs, navigational aids, wrecks and obstructions, prohibited areas, restricted access areas, anchorages, port plans, settings for safe depths, two different navigation perspectives, depth contours and coastlines, as well as tidal levels and more.

Thus, you may well have the widest range of maps to use on your GPS unit. You can therefore select the appropriate map display, according to your own taste.

Maps, however, are not mandatory for the operation of your Garmin device. The recording and representation of your travel information consists of self-contained data, which is not dependent on any map. Likewise, you can for example, load a track from an acquaintance, or from the Internet into the unit and independently follow the visible line in the display. Based on the representation of your own position on the display (blue position triangle), you just need to be careful to ensure you do not stray from the line.

Those who are well adept, have perhaps no problem with this situation. Certainly, it is a very great advantage if, for example you can see a nearly parallel diverging fork in advance, based on the underlying map display and determine if the line branches to the left, or to the right. In this case, without an underlying map, you only have a 50/50 chance to choose the right path.

Figure 1-2
Map view: Basemap with a loaded track and "Red Bike" waypoint

Either you would see on the screen that you are continuing to move on the line, or you discover after 50 m, or later, that it is the wrong way.

If only the basemap is installed in the unit, i.e. no other map with a higher accuracy, the unit's display remains fairly empty when you zoom to below the 2km (1.25 mi) representation. Attempting to start an automatic navigation to a selected waypoint would only result in the message:

"Routing error. No routable roads in this area..." The route will not display at all and in the best case scenario, display as a direct line (as soon as Setup > Routing > Activity "Direct Routing" is selected.)

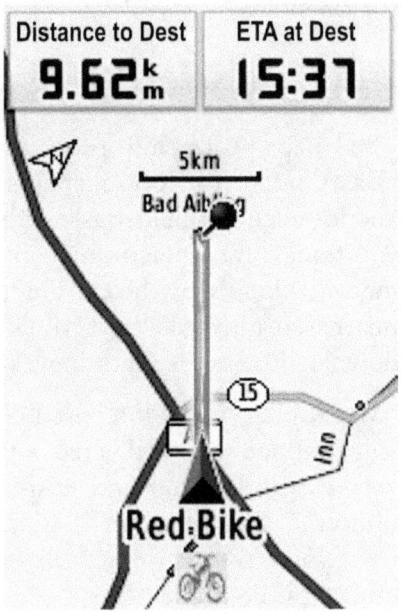

Figure 1-3
Direct routing shown on the Basemap

Routable map data from pre-programmed microSD cards are quickly ready for use. Insert the card into the device (in the battery compartment, under the battery) and you're off. Those who do not live in a border area of a map and use only one map, may well be quite happy with pre-programmed microSD cards. However, if you move around often in a border area, you would be constantly changing the SD card of one country with that of the other country, likely an intolerable situation.

With the DVD version of map data that is to be installed on a PC, one has the possibility to upload the map, or multiple maps, to the GPS device via the "Install Maps" function of the associated PC map Software, "BaseCamp." The respective maps could be sent as a single map, or as individual map tiles. Thus, you can work with the cross-region coverage area of the map in the GPS64: When you cross the map edge which ends, for example, at the German border, the next bordering map is immediately displayed, the adjoining Austrian map, for example.

Garmin Software: MapSource, BaseCamp, Connect

The map processing program "MapSource" (for road maps) or "BaseCamp" (for topo maps) is included with the Garmin map DVD and installed simultaneously with the map data on a home computer. If you aquire more Garmin maps on DVD, after installation they are managed jointly by the existing program. If you have map data from a purchased microSD card, you'll get the BaseCamp software as a free download from the Garmin website (see the beginning of the book).

"MapSource" is the long used GPS map software from Garmin, which will continue to be delivered with road map DVDs. It is well designed, very clearly laid out and enables basic planning and post-processing options. This program is no longer recommended to communicate with your '64 device, which is currently the 3rd generation of this outdoor GPS model.

"BaseCamp" was developed out of necessity in 2009 with the advent of the 2nd-generation GPS device, due to its new data structure. In the previous paragraphs, you'll have come across this name quite a few times, because BaseCamp has grown to become a very extensive GPS program, with which one can work much more flexibly and comfortably, especially when it comes to route planning, or post-processing. Besides BaseCamp giving you the detailed drawing and editing features in the 2D view and the graphic representation of terrain in the 3D view, you have the ability to virtually fly along the marked track (including those from the GPS device) and experience the many elevation details of the tour.

Likewise, compact excursion recommendations can be created with GPS data and photos and can be published as so-called "Garmin Adventures." On the other hand, people interested in electronic scavenger hunts can use BaseCamp to manage their geocaches.

With BaseCamp, it's up to you whether you want to work from the hard drive of your PC, or directly connected to the GPS64 device's storage. With BaseCamp, you can also use map data on the GPS64's internal storage, or its microSD card, exactly as if you had installed it from a DVD onto the PC. Since the Garmin map DVDs include a license for a single PC only, you can install the BaseCamp software on any number of other computers and use them to work with the maps from the GPS device, itself. On your next vacation, it would certainly be nice to be able to plan, or modify a couple hikes from your laptop.

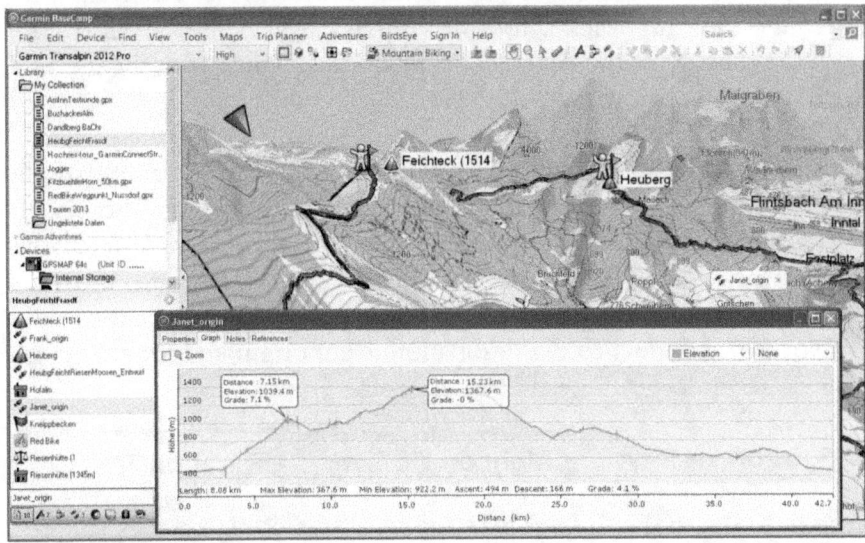

Figure 1-4 BaseCamp map software with Topo TransAlpin and a track with 3D view and elevation profile

Furthermore, the view of various sensor values such as temperature, pulse rate, cadence, etc., can be activated in the graphic representation, in addition to the contour lines of each recording.

Those who want to analyze their fitness levels to the smallest of detail, and also use the new live tracking feature of the '64s, can create a free user account with "Garmin Connect," the worldwide training portal for sports enthusiasts. If you lack appropriate maps to do the necessary holiday preparations, you can also draw your own new routes online, or search for other recordings. This online service will be discussed later more in-depth (Chapter 4 / "Evaluate track recording on the PC").

But faster than expected, we are already in the midst of the typical GPS technical terms. Therefore, as a first point, we must clear up what is what.

Routes and Tracks

Colloquially, everything was once referred as a route. After all, who says "We took the track across the Alps?" It was always called "We took the route over the Alps." For the understanding with the use of a GPS, we need to now establish a clear distinction.

As has just been mentioned several times, <u>routes</u> are the automatically calculated paths to a destination (also called auto routing). So exactly the same as we know it from an automobile navigation system, only without providing speech output. The '64 signals a turning situation with a beep, text message and graphic display. You can choose this type of navigation when the destination is important and the way to get there is secondary, or doesn't matter at all.

Figure 1-5

        map view | Active Route view | preview of a turn

A route serves the purpose of spontaneously calling up a destination in the GPS64 and to automatically calculate the path to it via the device. There are limited options to influence how the route is created (in the device: Main menu > Setup > Routing). Nevertheless, by selecting the "Activity," e.g. mountain biking, hiking, mountaineering, etc., as well as some detailed avoidance wishes (shortest distance, or ascent), the route can be adapted to your own ideas quite well.

However, if one has concrete ideas about which paths are to be used to a destination, or has a tour description on hand, then the navigation on the basis of a track is clearly a better decision.

A track is the point-by-point line of movement that the GPS device automatically records. A track can also be used for orientation of a repeated, or return trip by vehicle, or foot. In other words, it's the fabled "trail of breadcrumbs," that could have been created as if one had dropped individual breadcrumbs as path markers. These breadcrumbs create points (track points) that are automatically connected to each other and produce the line of movement. The track line in short: the track. A track of an 80km (50mi) mountain bike tour, for example, consists of about 3,000 track points (Setup > Tracks > Recording Interval: "Normal".)
Tracks are not only just the recordings of GPS devices, but can also be drawn in a GPS map on the PC.

If you want to use a track for navigation, you have to load it into the device prior to departure. One cannot produce a track in the unit before the tour, in order to find a specific target. A track is particularly suitable for navigation, if no routable map is available in the unit, when the tour runs as a circuit (typical of cycling), or if you just have a clear idea of which path should be used where the journey is the reward (e.g., hiking through the Alps).

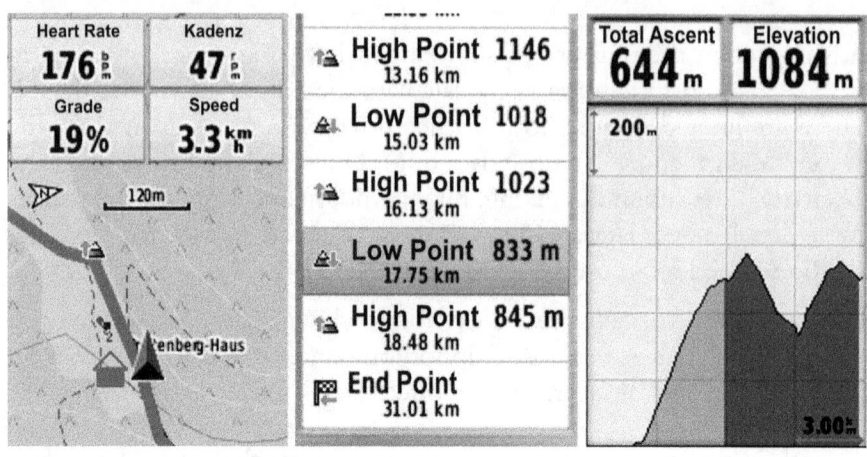

Figure 1-6 Screen display of a track
    Map view | Active Route view |Elevation Profile view
                                                                       (green = already recorded
                                                                       blue = upcoming)

→ Routes and tracks are two completely different types of navigation. In the BaseCamp mapping software on the PC, you can easily tell the difference between the two through the various symbols displayed before the name. However, in MapSource the various objects are managed in separate windows. ←

Track points, Waypoints, Via points and POIs

The points (breadcrumbs) that make up a track, are called track points. These are automatically set by the GPS device. If you draw a track on your PC, the mouse clicks are what generate these track points. When connected together, they form the track line, i.e. the track. You cannot attach additional information to track points.

Waypoints are specific points, which you can save to the GPS64 with the "MARK" key while in transit, for example, because a very nice view is found at this location. They are simply special points, which you want to additionally remember.

You can also create waypoints in the software on the PC, save it in GPX format and send it to the GPS device. This has the advantage that you can quickly select this waypoint in the unit and start automatically navigating to this point - routing. In any case, it will save you from the prolonged search in the destination entry options, or from typing in the address letter-by-letter. You can attach further comprehensive information to waypoints on the PC, such as a brief comment, links, photos, etc.

POI (Points of Interest) is an existing collection of interesting waypoints, and used to satisfy daily needs, or places to go in urgent situations.

These are included in the maps from Garmin, or can be downloaded, at no cost, as a POI collection, for example from the Garmin website, or from various portals, like www.poi.gps-data-team.com and www.pointoo.de.

Figure 1-7 FIND sub-menu; If the Topo TransAlpin map is on the device, mountain huts in the area are listed under Lodging > Destination.

In the GPS64 you call the POIs with the "FIND" key. This will open the destination menu with further subcategories such as hospitals, service stations, food and drink, shopping, entertainment, etc. You can also find all the other categories in the destination menu, which include navigation-type objects such as waypoints, saved tracks, routes and photos with GPS information.

1–25

On the PC in Garmin's BaseCamp, you search for POIs by typing the POI-term (e.g. Frasdorfer hut) into the "Search" field in the upper right corner of the window. However, alpine huts are only searchable with Garmin topographical maps.

Via points are found only when using routing. These are the mid-points that should be passed on the way to the goal, because you do not want to take a direct route.

Geocaching

For many, it's still hard to imagine that the treasure hunt we know from our childhood has an updated electronic version - geocaching. Grown-ups and children alike are using GPS equipment to search for small treasures in any type of environment. Here, the alleged treasures are rather symbolic. To be found is merely a weatherproof box with contents, or encrypted puzzles. The fascination can be quite different. One may be surprised by a little treasure while hiking with the family, when it's "only" about the guided search with the help of coordinates, while others go on an all-day hunt for caches. They don't have to only be permanent hiding places, but also short-term, temporary hiding places and from them, a unique speed event is created. It is also possible for anyone to design perplexing puzzles and give tasks to people he does not know, by combining several hiding places and tricky challenges with virtually unlimited difficulty. Additionally, coins, as an example, can be put in the hiding place, which the next finder takes, registers on the Internet and then should place in the next hiding spot. Thus, this coin can be watched, while it wanders around the earth.

In the devices of the new '64 series, 250,000 "traditional" geocaches are stored already. So you can immediately grab your device and start treasure hunting.

Track down a geocache in the field

Go outside and turn on the device. Wait for reception, meaning that the question mark icon in the map goes off. Press the "FIND" key and select "Geocaches" from the objectives categories that are shown. Navigate to the entry with the middle rocker switch and confirm the selection with the "ENTER" key. From the list of displayed geocaches, select any cache in your area. This will open a map view (3rd image, below) with the "Go" button at the bottom of the screen. You confirm the "Go" option with a press of the "ENTER" key and with that, start the navigation from your current position to the selected geocache.

Figure 1-8
Display the Geocaches in the area, with the "FIND" button

Or if you want information about the cache in advance, press the "MENU" key, while viewing the map. This opens an options menu where you select "Review Point" to find the details about the cache. Here, you will find information on the level of difficulty, the coordinates and the description from the owner, as well as comments from other geocachers who have already "lifted" the treasure.

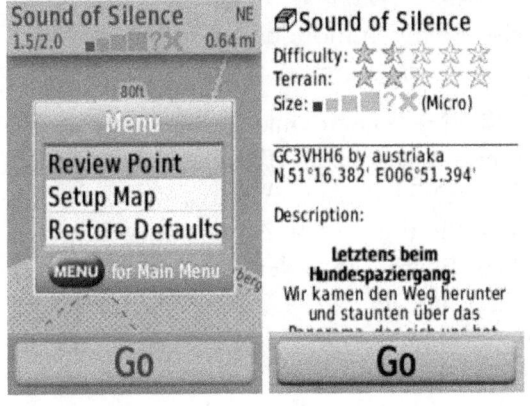

Figure 1-9
"MENU" > Review Point: See the characteristics of a cache

1–27

If there a question mark is displayed for the "Size:" indicator, then the size of the "treasure chest" to be searched is unknown and indicates a virtual cache. At the coordinates you don't need to look for a hidden plastic container, but must identify a specific code on an existing object there, e.g. an information board, which you must then re-use in some way, according to cache description.

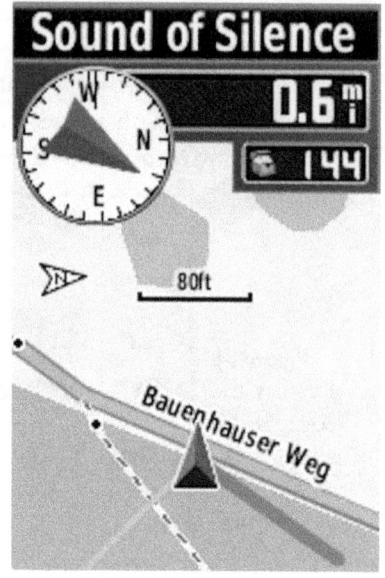

In the GPS64, there is already a profile prepared for use with geo-caching (Main Menu> "Profile Change"). Thus, you can choose your preferred settings specifically for this hobby, without having to constantly change the settings for other activities. For example, in this case, the additional "Geocaching" display in the map view is especially useful. Thus, the 3-axis electronic compass, the next cache with distance and direction indicator, and the number of caches already found, are visible at a glance in the map at the same time.

Figure 1-10 Map view with the extra "Geocaching" dashboard

The position coordinates can also be much more useful for geocaches, which may require a more intensive, local search. You decide!

To set up the map, press the "MENU" key (while in map view) and select "Setup Map" from the menu that appears. Then navigate to the "Data Fields" line and press the "ENTER" key. A list of options will open, with which you can change the top section of your map display as you desire. For example, "Dashboard," in order to conjure up the "Geocaching" details in the map, like in Fig. 1-10, or "Large Data Field" to display the position coordinates in UTM format in the next step. After confirming with "ENTER," return to the map with a press of the "QUIT" key.

Call up the options menu again with the "MENU" key and now select "Change Data Fields." The data field at the top of the map is now highlighted. Press the "ENTER" key to set the value that you want to see in the selected data field. In our case, that would be "Location (selected)." In order to be able to search for the hiding place in exact meters, you should also change to "UTM UPS" in Main Menu > Setup > "Position Format."

For the route calculation to the geocache, "Direct Routing" can even be beneficial, because you are immediately shown the direction directly to the cache on the display. This setting can be found in the Main Menu > Setup > Routing > in the "Activity" row.

In order to maintain an overview of the completed and outstanding hides, a cache that is found can be immediately "logged" in the device, i.e. marked as found. This requires that the navigation to this point is still active, and can be accomplished either in the map view with the "MENU" key and option "Found," or Main Menu > Geocaches > "Log Attempt." If it is a multi-cache (where success depends on detecting multiple hiding places), the option "Enter Next Stage," allows you to simply type in the position coordinates of the next hiding place.

Figure 1-11   Mark a geocache as "Found"

Your geocaches marked as "Found" in the device, can be called up via the "Geocache" category in the main menu. From the displayed list of geocaches, press the "MENU" key, then select the "Show Found" category. Maybe you would like to select this cache again, to which to navigate?

In order to not unnecessarily trample the surrounding terrain flat, when searching for treasures that are potentially no longer there, Garmin developed "chirp." This is a small transmitter which is attached to the cache container, so as to inform the seeker using an ANT+ compatible GPS device within a radius of 10 meters that the treasure still exists nearby. The chirp is password protected and can be freely programmed with coordinates and clues by the owner. In order to receive this signal, select the "chirp Setup" entry from Main Menu > Setup > Geocaches and turn on "chirp Searching." Programming of your own chirp is accomplished via the row beneath with the same name.

Geocaches: filtering in the device

In order to view a particular selection of the pre-installed geocaches on the GPS64, because at the beginning you would first like to start with a simple hide, for example, you can use the Quick Filter to specify the criteria for this search: From the Main Menu, navigate to "Geocaches" and confirm the selection with the "ENTER" key, in order to again display the list of all stored caches. Now press the "MENU" key and select "Choose Filter" > "Quick Filter." Navigate to the respective selection line, open the selections by pressing the "ENTER" key and select your desired options, which narrow down your search preferences the most.

Figure 1-12 Quick filter, in order to search for specific caches in the device

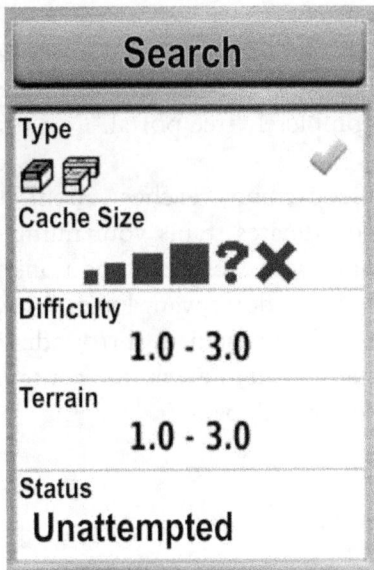

E.g. like in the image: Type: "Traditional and Multi-Cache," Difficulty of the find: "1.0-3.0," Terrain difficulty: "1.0-3.0," Status: "Unattempted, Did Not Find," The Type selection probably really only makes sense, if you have already loaded more caches in your '64, because the preloaded caches in the current version release are all only "Traditional" caches (with a hidden "plastic box"). At the end you start the search with "Apply Filters."

To search with the same attributes multiple times, you can also create search filters, which remain saved:
Open the Filter Setup selection in Main Menu > Setup > Geocaches and confirm on the "Create Filter" entry on the top line. Proceed in the same manner, as when using the quick filter and start the search with "Apply Filters." These newly created search options are then stored at once as "Filter 1, ... 2, ... 3" and can be selected and renamed again by entering the filter settings with "ENTER," and in the same manner, you can update the search criteria.

→ But beware: These specified search factors of an applied filter remain prevalent in their current state in the device. This means, when you display the list of geocaches that are saved in the device (in whatever manner, e.g. "FIND" key > "Geocaches"), only those that match the search factors of the last used filter are shown. In Setup > Geocaches, you can delete the created filter (scroll to the respective row with the rocker key, so that it is highlighted, press "ENTER" > "Delete" and again "ENTER".) ←

The collection of such geocaches can be found online on various geocache portals. For example, www.opencaching.com is a global and completely free portal.

Geocaching teaches about the use of GPS, and working with coordinates trains your number memory. A great thing to be used on days with weather conditions that don't lead to any big undertakings, but you don't want lull around, hobby free. You can still use it for fun in your immediate surroundings. Unfortunately, doing so in urban and metropolitan areas, the centralized crowd of seekers may inadvertently trample all parks.

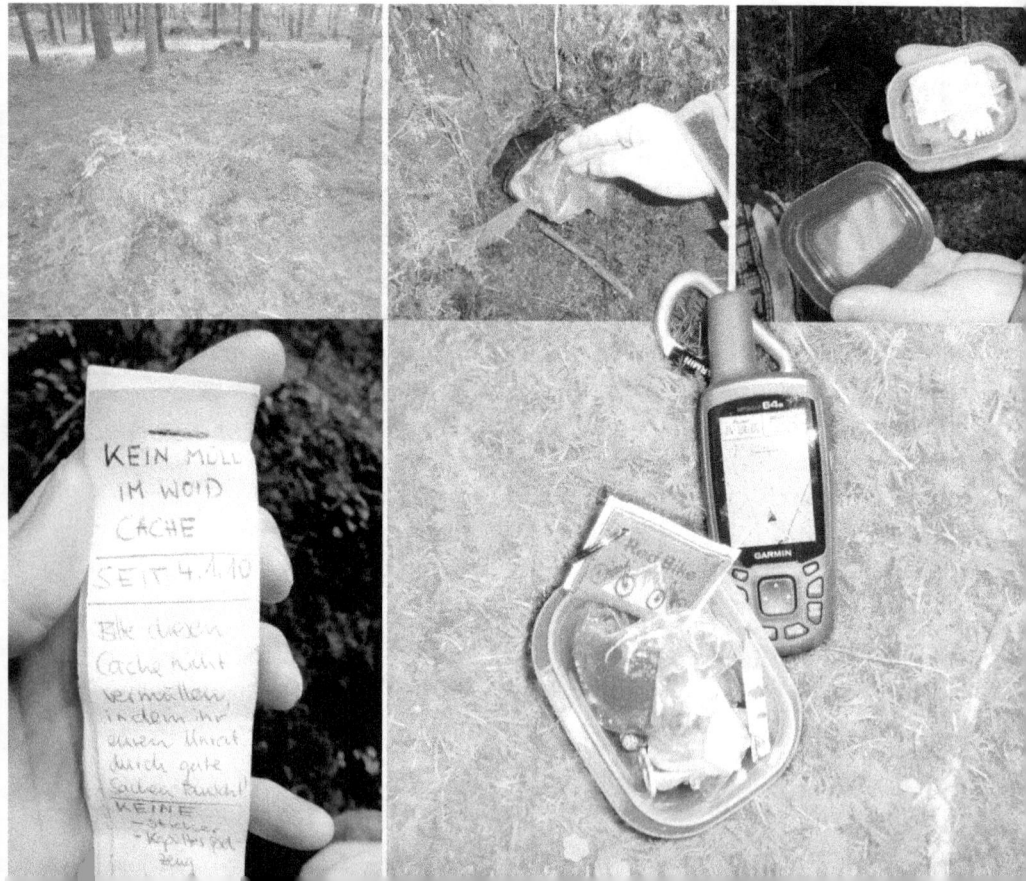

Photo tagging

Photo tagging is another way, based on the GPS navigation, to better remember adventures, or tell others about them.

GPS devices with integrated cameras and special cameras with GPS functionality, or connection options for GPS receivers, can automatically geotag photos, i.e. add the coordinates of the location. This data addition enables the location at which the photos were taken, to be represented along the track that was recorded by the GPS device, when viewed in the BaseCamp software on a PC. For this, in BaseCamp you click the track with the right mouse button and from the context menu, select "Geotag photos using track..." to enter the folder in which the external GPS camera's accompanying photos are located and generate the waypoints for them. These photos must match the recording time of the track. The photos are then waypoints on the track, which can be arbitrarily named and updated with further information.

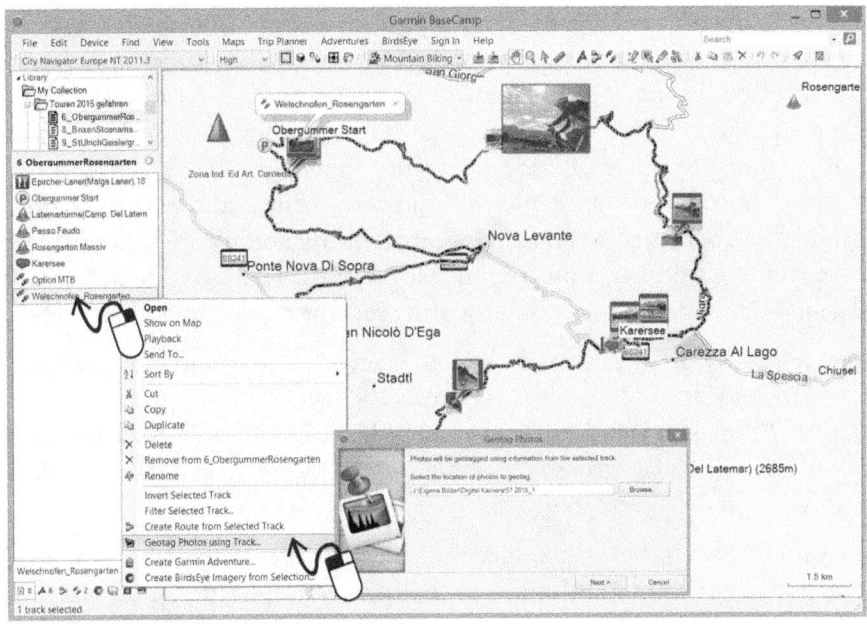

Figure 1-13   BaseCamp: georeferenced photos

In order to now eternally back up a track with the accompanying photos, you must use the Garmin-native file format "GDB" when saving. Only in this manner can you save all objects and shortcuts to photos in a single file, in order to be able to open it with the photos. In addition, the photos have to be saved in the same location.

But even as an owner of a conventional camera (without GPS), you can geotag your photos (see Chapter 4 "Georeferencing photos") and subsequently use these in your GPS64 for photo navigation.

Coordinate system

For GPS navigation, the map datum (reference system) WGS 84 and the map spheroid (ellipsoid) WGS 84 are used.

<u>WGS 84</u> (World Geodetic System 1984) is the geodetic basis of the GPS system, the measurement of the earth and its objects with NAVSTAR satellites.

To be able to designate a certain point on the earth, one needs a system which indicates the exact distance in latitude and longitude to a certain point. For this, a grid was placed over the earth (the coordinate grid). The equator, with 0°, became the starting point for reckoning in northern and southern latitudes and the meridian passing through the London borough of "Greenwich," the zero value and the term "Prime Meridian," became the starting point for reckoning in western and eastern longitudes. So now, the tiniest point on Earth can be accurately designated numerically, in other words, receive coordinates.

Yet even with this single system, there are numerous national networks/grids, such as the German Gauss Krueger grid with the datum "Potsdam" and the ellipsoid "Bessel1841", the Austrian grid with the reference system "Austria" and "Bessel1841", the Swedish grid with "RT90," and many more.

In order for a global agreement to be possible, emergency services, police departments, fire brigades, civil protection agencies, other relief organizations, as well as surveyors themselves, all work with the UTM coordinate grid with the geodetic datum and reference system, WGS 84.

The UTM coordinate system (Universal Transverse Mercator) was developed in 1947 by the armed forces of the United States. In the context of internationalization, it displaces individual national coordinate systems more and more. So one day, the official German topographic maps using the Gauss-Krüger coordinate system will probably have been entirely replaced by the UTM coordinate system, due to the WGS 84 reference ellipsoid.

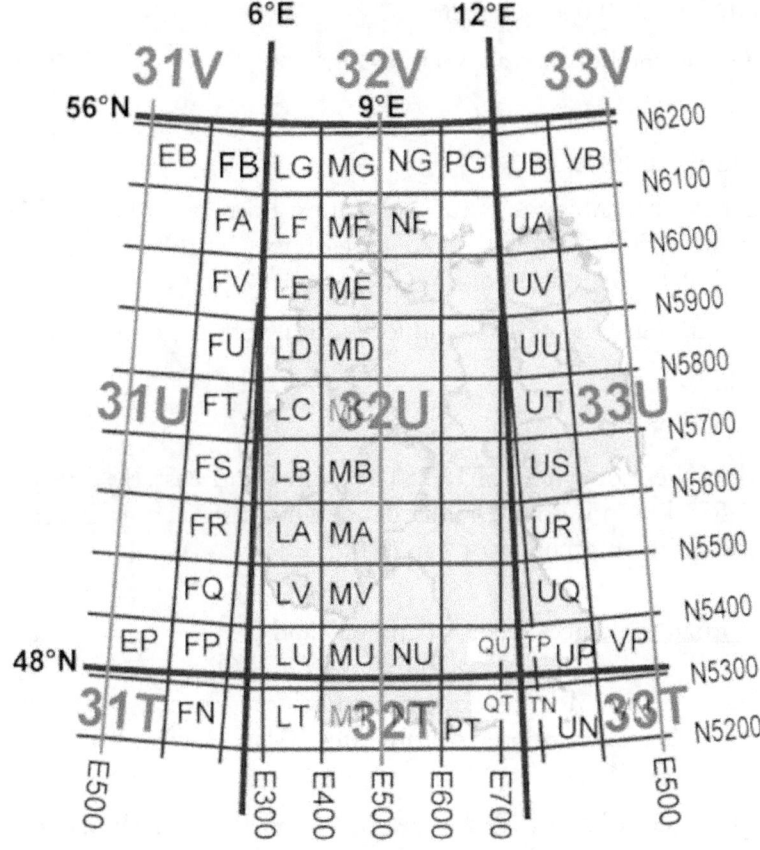

Figure 1-14 Source: WIKIPEDIA UTM-Zone fields example, Germany

The representation of coordinates in UTM format is termed as a square grid orientation. This is expressed in meters. The format always begins with a one- or two-digit number, followed by a letter which represents the UTM zone.

The upper row of numbers which follow, gives the measurement for the east-west position within the zone in meters. This value is therefore called "easting."

The lower row of numbers gives the measurement for the north-south position from the equator in meters. This is the "northing." In order to avoid negative positions in the southern hemisphere, the equator is assigned the value 10 000 000 meters.

Note: "Walk in the house, walk up the stairs"

Example position format:

GPS64 Series Main Menu > Setup > Position Format > "UTM UPS" (UPS for the system used in the region of the Earth's poles). Map datum: WGS 84 / Map Spheroid: WGS 84.

The waypoint: Red Bike, 83131 Nußdorf a. Inn, with the coordinates

33 T 0285494
     5291575

is located in the Zone field 33T, 285,494 km in an easterly direction and 5,291,575 km in a northerly direction. The position can therefore be described with an accuracy of 1 m (3 ft.).

Position format in degrees (a.k.a. "Latitude" and
                 "Longitude" = LAT/LON:
GPS64-Series Main Menu > Setup > Position Format > "hddd° mm.mmm' " (Degrees and decimal minutes), Map Datum: WGS 84 / Map Spheroid: WGS 84.

The waypoint coordinates: Red Bike in Nußdorf a. Inn should now read

N 47°44.511'
E 12°08.313'

Other delineation formats for these coordinates can also be selected:

- hddd°mm'ss.s" = Degrees Minutes Seconds

or

- hddd.ddddd° = Degrees only

➔ Note: Once the map datum and map spheroid WGS 84 are set in the settings, the GPS navigation runs correctly with Garmin maps. ⬅

The format of the representation, whether in meters or degrees, is at the discretion of the device's user. The deciding point is most probably the further processing of waypoints with various GPS programs on the computer, where you prefer your particular setting, or it is used to match up with the hiking map used. Often even 2 grid notations are printed on paper maps.

Caution: in conjunction with the coordinate system used in the hiking map, different map reference systems may have been used, such as RT90, Rome1940, Potsdam, NAD27... This can be seen in the legend of the map and accordingly adjusted in the GPS64, if you want to synchronize the device to this map. Newer maps, as well as all Garmin maps, use WGS 84.

North reference / Declination

The constantly changing magnetic field of the earth and the resulting deviation between the magnetic and geographic North Pole, must be considered when reading maps. This would be set on a compass, if working with one. On paper maps, one finds this deviation information in the legend. This is classified in the deviation of the map itself, and in the compass.

If you're using a combination of compass and GPS, you'll be happy to be able to set the north reference GPS64 to "Magnetic" to match the compass. When working with a combination of paper maps and GPS, you would choose the north reference option, "Grid."

If you work only with the GPS, you don't have to worry at all about the magnetic variation, or the grid lines. Normally, you would just use the true north reference:

GPS64 Main Menu > Set Up > Heading > North Reference > "True"

GPS + GLONASS

The Global Positioning System GLONASS, which is operated and financed by the Ministry of Defense of the Russian Federation, is similar to the US NAVSTAR GPS, the GPS system currently in use. Developed since 1972, the Russian system has been in use with full global coverage, since 2012. The components of the new '64 series are GLONASS-capable.

The receiver of the '64 series indicates the reception of GLONASS satellites, as soon as it is selected in the device Setup > System > Satellite System > GPS "GPS + GLONASS" (default). With this, you have an option, should you notice long-lasting, wide variations of your position arrow in the display, as compared against your real position. If this is the case, switch to the conventional "GPS."

WAAS und EGNOS

For a higher accuracy of position determination in landing approaches, the WAAS (Wide Area Augmentation System) system was developed for the American aviation authority. The WAAS data increased the accuracy of the GPS signal only in North America, as the correction data is calculated and transmitted only for this region. The reception of the WAAS signal is still partially possible in Europe, but can lead to inaccuracies in positioning, when used there.

The European service for improving positional accuracy is called EGNOS (European Geostationary Overlay Service). With EGNOS, the accuracy of the GPS system in Europe is expected to increase to 5 m (15 ft.). EGNOS consists of a network of ground stations which receive GPS signals and transmit them to the so-called "Master Control Center - MCC". Like WAAS, corrections to the deviation of the satellite clocks is being sent. Drift in the satellite clock signals come about due to influences of the atmosphere and ionosphere on the signals.

For land and sea navigation, the accuracy achievable with WAAS / EGNOS is usually not necessary, because the accuracy of the map representation is clearly worse.

Updates

Technical development races ahead quickly. Do not be afraid, the unit may have already been out of date, when you bought it.

The software in the device and for making plans on a PC will be kept up-to-date by Garmin for free.

At software.garmin.com, you can find "Garmin Express." This is a tool that needs to be installed on a PC. With a GPS device connected, it will search for the latest device software, without additional effort.

All Garmin programs installed on the PC update themselves automatically, when the computer has Internet access and nothing else was chosen in the settings options (in the BaseCamp menu bar: Edit > Options, "Software Update" tab). Otherwise, the update process can also be started manually (menu: Help > "Check for Software Updates ...").

Chapter 2 - The Device

Now that you know about all the kinds of adventures you can undertake with your GPS64, which maps you may still need, how to spontaneously choose a destination and let the device automatically calculate the way there, or rather stay at home and follow a mapped out tour on the PC... Once you know which map datum must be set and how you keep your device up to date technically, then you're almost ready to go!

Get to know your device by practicing with changing typical settings. We do not want to chew through all the different settings variants. What you can do with other settings, is listed quite well in the owner's manual section titled, "Customizing Your Device." This can be found online: www.garmin.com/manuals, or when you connect your device to the PC, launch the "Garmin Express" tool on the PC and click your device on the index page. There is a "Tools" button, where you can find a link to download the operating manual.

Inserting batteries and turning on the device

Well, get that awesome thing out of the package! Insert 2 AA, or rechargeable batteries. The longest operating times can be obtained with charged NiMH, or lithium batteries, the latter are recommended for frosty days. A set of batteries should supply the GPS64 with enough power to last for 16 hours, under normal conditions. Charging via the PC cable, or other charging cable, can only be done when using the optional Garmin NiMH battery packs (Garmin Art no. 010-11874-00).

The battery compartment can be opened by lifting the lid of the rear panel (turn the D-ring anticlockwise and lift), where you will also find the serial number and the slot for the microSD card. Above the battery compartment, held with a rubber cover, is the slot for the Mini-USB port. The GPSMAP64s and ...st models also have an MCX connector for an external GPS antenna that enables improved reception in car interiors.

The keys and their meaning

The only button found on the side of the device is the "POWER" button on the right side. A short press will turn the power on and with 3-seconds of pressure, off again. A short press on Power, when the unit is on, will switch it to the status display, where you can turn on, or change the brightness of the backlight, read the battery level, the quality of GPS reception, the Bluetooth activity, and date and time. Here, with each press of the "POWER" button, you can change the level of the backlight between bright, medium, or off. After 8 seconds of no action, this display is automatically hidden. This light setting persists, even after turning the device off and on, again.

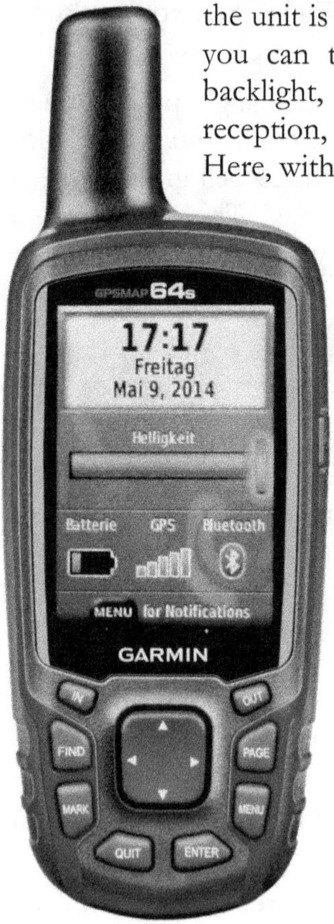

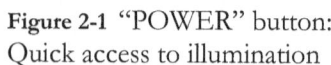

When outdoors, you will hardly need, or even perceive the illumination, as the display perfectly exploits the reflection of sunlight for a high-contrast "lighting." Don't forget to turn off the light with a short press of the on/off key, especially when the Backlight Timeout is set to "Stay On" (Setup > Display).

Figure 2-1 "POWER" button: Quick access to illumination

With the "QUIT" key (on the left under the rocker), you can also exit this view again. This key is also used whenever you want to stop, or cancel a called-up menu. If no menu is called, this key can also be used to scroll through the page sequence, but in the reverse direction (because it functions as the "back" key) as the "PAGE" key (to the right of the rocker key). This key scrolls through the preset page order to the next adjacent index to the right of that, which is currently displayed symbolically in the center of the circular page sequence. Each keypress rotates the page sequence one space to the right and opens the selection displayed in the middle after about 2 seconds. The selection of fields to be displayed and their order can be arranged in the Setup > Page Sequence.

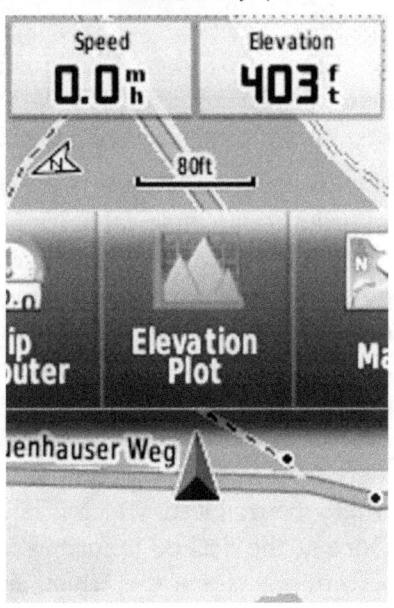

Figure 2-2 With PAGE, move through the sequence

Additionally, with one touch, the "PAGE" key can also be used to return to the Main Menu, when inside of a widely ramified submenu (e.g. adjusting the position format).

The "MENU" key, located just below (bottom right of the rocker), calls the Main Menu from any screen view, with a double-press of the key (2x, in quick succession).

However, a single press opens the options menu which exists for the current view. Please keep this key in mind, in case you are in a display, and don't know your way around. Often one does not think of this key when they, for example, are located in the Waypoint Manager and wonder: "How can I delete the waypoint?" Then press "MENU" once and see which tasks are available. One often expects no further options such as in the "FIND" > POIs selection. If you are in the subcategory "Cities," the neighboring cities in the area are listed. This often includes only a range of 10 km. But in order to get to the city outside of the 10km radius, for which you are now looking, you can use the

"MENU" key to call the "Spell Search" option and press the "ENTER" key to confirm. A keypad opens and then you can enter the city literally.

→ Do not forget: You don't know where to begin, with what is being displayed? Ask the "MENU" key which options it provides. At first, this is often very surprising. ←

If you want, here's a test using the example of the map view:

In order to target an arbitrary point in the map, to start a navigation to get there, or even to learn advanced information about this point, first press the rocker key on any page, on which a pointer is shown on the map display. This movement will continue, until you release the rocker key. In this manner, you can navigate with the arrow pointer to an object in the map that interests you. If you bring your pointer to the edge of the display, the map will slide further and you can move the map view in that direction (for example, perhaps you want to know where the trail lead, which is being "cut off" at the edge of the display). Once at the desired point, press the "ENTER" key. At the top of the screen, several choices which are stored for this point will then appear (Figure 2-3, 1st image from the left).

Suppose that we want to highlight the point directly on the road. So let's highlight the "path" (otherwise select another entry with the rocker key) and confirm the selection by pressing the "ENTER" key. In response, detailed information about this point opens (2nd image from the left). Now you can start an automatic navigation to this point, by using the "Go" selection that appears at the bottom of the screen. But first, we want to save the item for later use. To do this, press the "MENU" key and confirm the only choice that appears, i.e. "Save as Waypoint," by pressing the "ENTER" key (3rd image from left). After the successful appearance of the "...saved..." message, when you view the details of this point, you will now see the Waypoint symbol in the row with the name of the point. This Waypoint is now also saved with the flag icon in your GPS64 (4th image from the left).

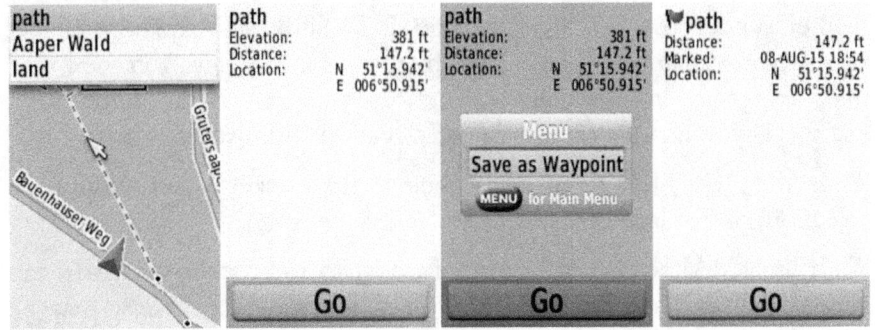

Figure 2-3 In the map view, target the point with the rocker key, display with "ENTER" and save with the "MENU" key

With a press of the "QUIT" key, you would now return back to the map view, where you should see the flag. Before you do that, you can also press the "MENU" key anew, for further waypoint actions, e.g. edit the details for this waypoint. After pressing the "MENU" key, confirm the only option that appears ("Edit Waypoint"), where detailed information appears, broken down into individual boxes (middle image).

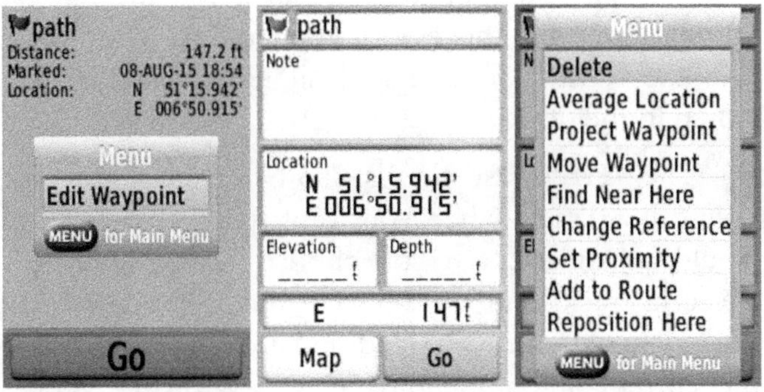

Figure 2-4
Delete Waypoint

With the rocker key, you can now jump to the appropriate box, confirm with the "ENTER" key and then change the entry, or the symbol. On the other hand, if you would you like to "Delete" this Waypoint from your GPS64, press the "MENU" key again. Now, a very long list with a number of options opens, giving you choices of what you can do with your point (3rd image from left). At the very top, you'll find the "Delete" option, which you then must confirm only with "ENTER."

2–45

In the example above, we witnessed the function of the rocker key (center, under the display) live and it is fairly self-explanatory. Of course you can move the map with it, as well the elevation view and make selections in all menus and options fields. In the elevation view, use this key to change the display of elevation and distance.

With the "ENTER" key (right below the rocker), you confirm a selection, or displayed message.

With the "MARK" key (bottom-left, next to the rocker key), you can save points that are important to you, while underway. These may be named exactly in the device, or edited afterwards at home on the PC as a single Waypoint and stored.

The primary purpose of the "IN" and "OUT" zoom keys (directly below the display) is to reduce and enlarge the map and Elevation Plot view. However, in long selection lists, you can use them to scroll up, or down one page at a time, such as from the Main Menu, where just using the rocker key would only scroll line-by-line

And finally you have the "FIND" key remaining (to the left of the rocker key) as the main key for the starting/stopping of all navigation tasks. When you want to start an automatic navigation (Route) to a destination, press this key to enter the selection menu, where you will find the POI collection and all navigation-suitable objects specially stored in the device.

→ The "Tracks" section is also here, although you do not necessarily only call a Track, when you want to start navigation. With a track, it would be sufficient if you choose Main Menu > Track Manager > (select a track) and then select "Show On Map" and "Set Color." A track doesn't have to be additionally calculated, unless you want to see the current and impending elevation profile. You will only see the latter, when you start navigating the track with "Go." ←

With "FIND," you also stop all started navigations. If the goal was not automatically detected, or not at all reached by you, you must again turn off the active route itself. This time this is not done with "QUIT," but with the "FIND" key, just as you did the previous navigation tasks.

Overview of keys and key shortcuts

POWER - Short press - Long press, ca. 3 sec	right-upper side of the device = Switch on the device = Illumination, status page = Gradient illumination = Switch off the device

IN + OUT zoom keys
- Short press
 	
- Long press
	directly under the display
	= Gradual change of the
	   map scale and the
	   elevation profile
	= Page scrolling in menu lists
	= Fast zooming

PAGE - Short press - Short press, twice	right of the rocker key = Scroll right through the main pages = Return to the main menu from a a widely branched subcategory = Return to the last application
MENU - Short press, twice - Short press, once	right-bottom of the rocker key = Main Menu = shows the options menu for the current view
ENTER - Short press	bottom-right of the rocker key = confirms a selection

2–47

Rocker key - Move in all directions - Move up & down, then right & left - Move right and left	middle, under the display = Navigate the pointer in the Map, navigate menus and drop-down lists; = Change elevation profile scale = Show specific position in the Elevation Plot view
QUIT - Short press	bottom-left of the rocker key = Exit, or cancel open menus; = Scroll right through the main pages
MARK - Short press	right-bottom of the rocker key = Save the current position
FIND - Short press	right of the rocker key = Open the destination drop-down menus; = Starting and stopping automatic navigation
Key combinations for troubleshooting	
PAGE + ENTER - simultaneously press and hold, in addition to pressing POWER to turn on the device. Keep PAGE & ENTER pressed, until the warning message with the question appears.	= Hard reset: the GPSMAP 64 is reset to its factory settings, all user data and settings are lost (confirm with "Yes"). Thus, any malfunction can be eliminated.

Working in the map view

After switching on the device, the map display is opened. The blue triangle in the lower map area displays your own current position. If you move with your GPS64, the map moves along with your <u>motion</u>.

→ However, there is also a case where the map no longer follows along with your movement: If you happened to have pressed the rocker key to target the pointer to a position on the map, or the map has shifted for the purpose of displaying information - as long as you can see the white arrow in the map - the map orientation will remain as it is and does not follow along with your movement. Press and release the "QUIT" key to return to the normal map mode. ←

Figure 2-5
Map view in manual mode; that is, the map does not move along with your own movement. Go back to normal map view unconditionally, with a press of the "QUIT" key!

Working in the Elevation Plot view

Press the "PAGE" key repeatedly until the "Elevation Plot" item is visible in the middle of the Page Sequence display. After about 2 seconds, the view opens automatically. (If the Elevation Plot item cannot be found in the Page Sequence, call up the Main Menu by pressing the "MENU" key twice. With the rocker key, navigate to the "Elevation Plot" and open this view by confirming with the "ENTER" key.)

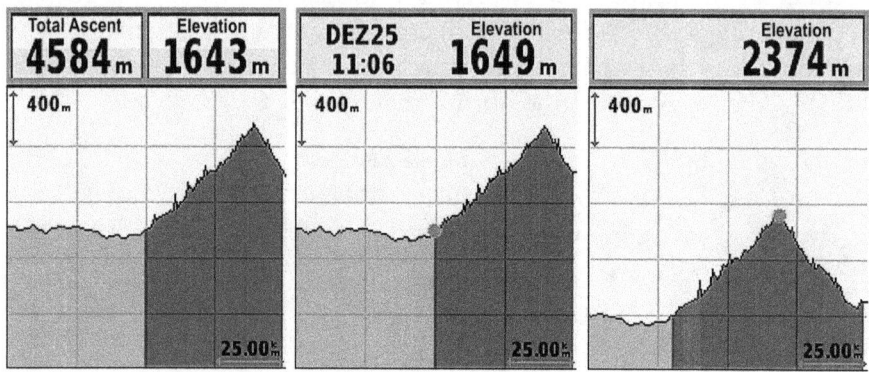

Figure 2-6 Elevation Plot view: green = previously recorded elevation profile, blue = forthcoming elevation profile

In the figure, you see the elevation of the course. One half is blue, the other green. This is the case if you have already recorded a certain distance with your GPS64 and additionally, a routing is active, i.e. an automatic navigation was started. One way this can be achieved in the GPS64 is by selecting a destination. Another way is by activating the "Go" button of a saved track. The green part of the Elevation Plot is your previously conquered ascents and descents. The blue part shows the profile of the elevation, which is still pending. So if you have neither selected a target point in the device, nor have you started a track, or a route, you will see only the green, currently recording elevation profile.

In the upper part of the graph, you will find two data fields, where, per factory settings, you can read the heretofore Total Ascent and current Elevation. (With a press of the "MENU" key, the information within these data fields can also be changed.) Pressing the rocker key on the left, or right side, will display a red position ball on the contour line. The exact information of this position can be read in detail at the top of the screen. With a press of "QUIT," you will return back to the normal Elevation Plot view (position without ball).

On the other hand, pressing the top, or bottom of the rocker key, puts you into the Elevation Plot's adjustment mode. At the top of the screen, the message "Adjust Zoom Ranges" appears. With a press of the upper/lower portion of the rocker key, you can change the display's altitude scale, by pressing the left/right side you can change the distance. When underway, it is normal for you to change this display scale frequently, in order to best estimate how far it is to the next summit.

With a press of the "QUIT" key, you exit the adjustment mode, or with a double press of the "PAGE" key, go immediately back to the last application/navigational page view.

Basic Settings

...are something you no longer need to set. In a user-friendly manner at the initial turning-on of the device, the '64 asks for the language to be used. For example, if you were to select "German," based on this selection, all basic settings would initially be adjusted to German usage. This means, for example, metric units, the temperature in degrees Celsius, altitude in meters, the time format will be displayed as "24 hours," etc., and the time zone will be determined automatically based on the GPS position. Subsequently, you can also change all these things again through the Main Menu > Setup > "System" (Language) "Units" and "Time."

After switching the GPS64 on, it always starts searching for the satellite GPS signals. For this, it is advisable to move to an open area, with an unobstructed view of the sky. During the satellite search a small question mark flashes on the position arrow in the map view. If no signal is found, a message is displayed asking how to proceed. If you are in an area and don't intend to have GPS reception, select the "Demo Mode," which stops the GPS reception. If the search was successful, however, the position arrow jumps to your real position and the question mark disappears. An accurate representation of the current reception quality can be found in the Main Menu, by calling the "Satellite" category.

For our ensuing adjustment and exploration purposes and to save power, we can turn off the GPS receiver immediately. In order to do this, open the "Setup" icon in the Main Menu and then sub-category "System."

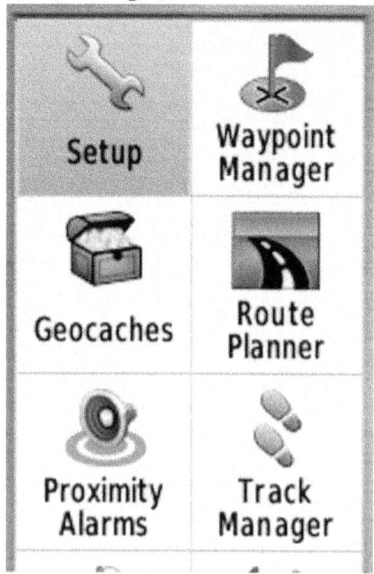

Figure 2-7 Main Menu

Since the 1st line, "GPS," is already highlighted in color, and is what we want, you only need to press the "ENTER" key now, to open the available choices. Select "Demo Mode" to disable the senseless indoors GPS search. Confirm your selection again by pressing the "ENTER" key. The selection is accepted and closes. (Please note: if you turn your GPS64 off and on again, this setting will be reset to normal GPS reception. So, don't be worried that you'll have forgotten this setting.)

It is also good to know the setting in line Battery Type: Here it is useful to select the type that you have inserted, in order to ensure the best type of power supplied. Conventional batteries are not charged when you connect the device via USB to the PC or AC power. The only exception is the optional Garmin NiMH battery pack. This is charged in the device. Using these batteries also eliminates the possibility of the battery type selection, since they are automatically detected.

The Interface setting should remain set to "Garmin Serial," if your GPS device is to be recognized as an external drive, in order to transfer all GPS data to, or from a PC, as well as to allow BaseCamp access to the '64. The options "NMEA ...", "RTCM" and "... Spanner" are provided, among others, for the connection with maritime equipment. You can also use the "... Spanner" mode, when the device is connected to an external power supply via the USB cable, but should still remain in normal GPS operation mode.

With one press of the "QUIT" key, you return to the Setup menu. We'll now skip some adjustment categories, such as the Display and Tones. These are optimally set at the factory and almost exclusively dependent on the tastes of each individual.

Beginning with the "**Map**" category, we'll slow down our pace and enter this sub-menu by selecting it with the rocker key and confirming the choice by pressing the "ENTER" key. Here you will find the settings that simplify the handling of the device while navigating with map:

- The <u>Orientation</u> is set to "Track Up," which you will especially appreciate, while in motion. In order to orient yourself to the state of the map, it may occasionally be helpful to set the map to "North Up" here. Those who may not want miss out on the three-dimensional view of a typical car navigation system, can choose "Automotive Mode" here.

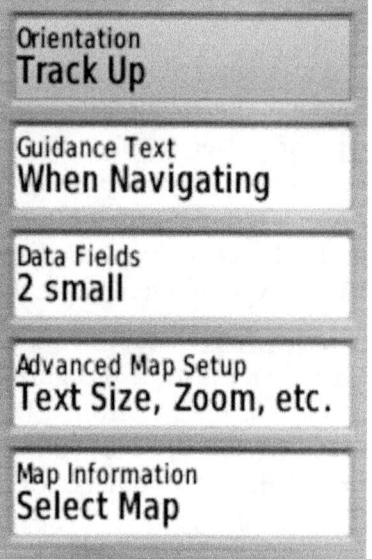

Figure 2-8
Setup > Map

- Leave the <u>navigation (Guidance) text</u> set to display only "When Navigating," in order that the entire map can be seen in all other cases.
- In the <u>Data Fields</u> box you can display additional information in which you are constantly interested, without the extra step of leaving the map view. You can choose from up to 4 data fields, or display an already available Dashboard, such as Elevation Plot, Stopwatch, an Automotive tachometer, Compass, Geocaching, etc.
- In the <u>Map Information</u> field, the maps can be selected or deselected, which you want to see in the map view, or are to be used by the GPS64 for route calculation. (Assuming you have several routable maps with the same coverage area loaded in the device, or on the microSD card. The visibility of topographic maps lies on top of road maps. If the route activity should be set

to travel by vehicle, you are advised to disable the topographical map. Always leave the Basemap active, without fail! This allows fast screen refresh when zooming out outside of the 20km [12.5mi] scale.)

The **Advanced Map Setup** field, listed in the second-to-last row, allows further map settings which affect:

- The Auto Zoom function, which automatically and optimally enlarges the display size that represents the view of your next turn. Choose "Off" if you find this to be disturbing;
- The Zoom Levels, which cause the map display to automatically increase in size, upon reaching various points on the map (assuming: Auto Zoom "On");
- The Text Size of objects in the map. You will certainly want to have User Waypoints (your own) displayed larger than other map information;
- The level of Detail to be displayed on the map (correspondingly, the speed of map image loading is also affected);
- The display of the Shaded Relief map, which is better left set to "Auto"matic, lets you better see 3D terrain shading in a zoomed-out map, while the detailed presentation of scale, e.g. "120m (400ft)," gives the best contrast-rich map view, without this 3D shading.

➜ With the "MENU" key, the settings can be reset to the original factory settings, meaning only those found in the field with which you are working. ⬅

Now leave the map settings by pressing the "QUIT" key twice.

As discussed in Chapter 1/"Coordinate system," in the settings for **"Position Format,"** you can choose between multiple representations of the coordinates per your preferences, which refer to WGS 84 in the Map Datum and Map Spheroid.

For a clear and meter-precise communication, the position specified in the UTM format is of great importance. By using the last number, you can give a reading down to the exact meter.

On the other hand, the representation format in longitude and latitude (hddd°mm.mmm') are clearly more popular, when it comes to dissemination on the World Wide Web, because you can immediately assess, on the basis of the degree, an indication of where the point is located on the globe.

Figure 2-9 Setup > Position Format

➔ However, unless you want to align your GPS to a paper map, which uses a different map datum and another datum point, Map Datum (datum) and Map Spheroid (datum point) must always have "WGS 84" selected during GPS navigation. ←

With a press of the "QUIT" key, you return to the "Setup" menu.

In the next "Setup" submenu, **"Heading,"** you can select all the options that concern the compass. Open this. In the top-most field, "Display," you can change the division scale of the compass, in case you want to have the graduation displayed as directional letters rather than with the specified angle (360°), or as nautical strokes (6400 mils).

The "North Reference" is correctly set at the factory to "True." Only if you work with a map, or compass, and want to calibrate the GPS64 to this, the corresponding settings "Magnetic" for the alignment of the compass and "Grid" to match the map will need to be changed, see Chapter 1 / "North reference"

The Go To Line (Pointer) offers the Course (CDI) mode, which allows you to not only see the compass direction to your destination, but also the current deviation from the proposed course (useful when using "Sight 'N Go").

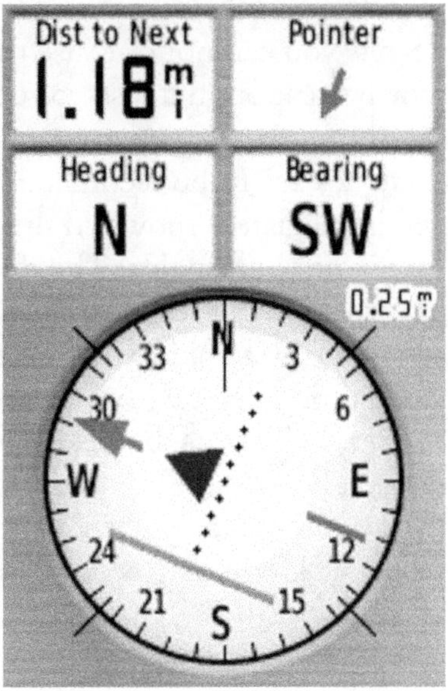

Figure 2-10
Compass: CDI-Mode for displaying targeted course and the ratio of the current deviation;
Division scale: "Directional Letters"

With the "Auto" setting in the Compass field, the compass mode automatically switches from normal compass operation to the GPS compass operation, when you are traveling at a higher speed, or you can switch it off completely.

You should do a compass calibration whenever you change the batteries, after moving long distances, or extreme temperature changes have occurred. Items that affect magnetic fields, such as cars, buildings, or overhead power lines, can affect the compass.

-Short Break-
The basic settings for general navigation tasks have now been touched upon.

Let's take a look at

Other meaningful settings

Since you can not only use the GPS64 for navigation, but also to record your own movement, you can decide which activites are to be recorded. These can be immediately shown in the map in a color of your choice. The decision, as to whether the recording is active and whether is should be displayed on the mapn or not, can be carried out in Setup > **Tracks** by selecting "Track Log." For example, choose "Record, Show On Map," when you start the tour - "Do Not Record," when you finish your tour.

Figure 2-11
Track settings for recording a tour

Track Log
Record, Show On Map

Record Method
Auto

Recording Interval
Normal

Auto Archive
When Full

Color

You can also do the following in the Track settings:

- Set the Record Method, that is, whether the track points are to be set at a given time interval, or a certain distance between each other, or automatically (more often in curves, less often on straight lines);
- Increase, or decrease the Recording Interval (the frequency of setting the track points);
- Set Auto Archive of the current recording to daily, weekly, or only when the current track is full with 10,000 track points. The track will archived separately in the device storage, which will accommodate 200 tracks. The setting, "When Full," makes the best use of the archival storage. "Daily," or "Weekly" can provide a better overview and less rework on the PC

and

- Adjust the Color with which the track log is to be displayed in the map view.

Here, the preset settings of Record Method set to "Auto" and the Recording Interval to "Normal" are usually quite perfect.

Exit this submenu with the "QUIT" key and in the "Setup" menu look at the next point "**Reset**." In the "Reset Trip Data" field of the submenu that opens, you have the option to set the data fields on the trip computer to zero. This is important to do, in order to only read the real trip data (length, duration, elevation...) of the current tour.

In the "Reset" submenu, one has the additional possibility to

1. Delete All Waypoints and to learn the percentage of Storage Used (how much storage is being taken by your waypoints, to see if you should possibly throw out some of the no longer needed waypoints),

2. Clear the Current Track and read its storage utilization,

3. Clear Track and Trip Data simultaneously resets the trip data, as well as deletes the track recording that is currently in the track memory. In other words, a "Reset" and point 2.

and

4. Reset the device to factory settings via "Reset All Settings" (i.e. to remove all personal settings, such as the arrangement of data fields, profile and routing settings, etc.).

All actions to reset, or delete the recorded data can be done faster
- in the Trip Computer view,
- in the Waypoint Manager

or
- in the Track Manager

with the help of the "MENU" key.

And as usual, return back to the Setup menu again with the QUIT key.

By calling the "**Page Sequence**" category in the Setup menu, you can place the applications, which you constantly access, into the list as you desire, which you then rotate through quickly with the "PAGE," or "QUIT" button.

For example, if you're timing a drive, the Stopwatch application comes in handy. From the map view, you can switch to the "Stopwatch" with a press of the "PAGE" key and trigger accumulated and interval timings, as well as stop, or reset them.

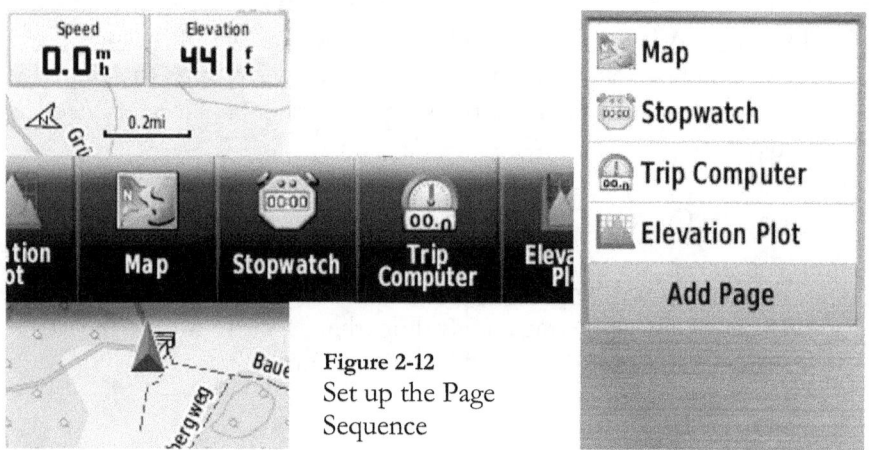

Figure 2-12
Set up the Page Sequence

For those of you who often start a navigation, the "Active Route" page would be interesting to have in the Page Sequence, because that is the list of upcoming turns of a route, or the highest and lowest points of a track. In this way, you can have advanced information, about what comes your way next.

If you set up too many pages here, you will struggle to quickly scroll to a specific view.

If you have added, or deleted your desired pages, return back to the "Setup" menu with "QUIT."

By now you have realized that your GPS64 can be customizable at almost all levels. So of course, the arrangement of the application icons in the **Main Menu** is also possible. During your first outing, pay attention to the features (applications) you access constantly and arrange them accordingly for faster access. While in the Main Menu, simply press "MENU" key (Fig. 2-13, 1st image from left) and confirm the "Change Item Order" selection, which appears, by pressing the "ENTER" key. A menu appears with all the selectable categories available in the GPS64 (2nd image from left), with the exception of those which already arranged in the Page Sequence ("PAGE" key). With the rocker key, navigate to the line for the item you want to change in the order and confirm the selection with "ENTER." The options "Move" and "Remove" then appear (3rd image from left). After selecting "Move," the line will now move along with the motion of the rocker key and will be dropped again, only if you press "ENTER." At this position, the item you just moved will remain.

Figure 2-13 Reordering the items in the Main Menu

However, if you have chosen "Remove" at some point, the additional option of "Add Page" is available at the very bottom of the category list, which lets you again reactivate the hidden categories and then subsequently move them back to their position.

Are all category elements in place? Then return back to the Main Menu with the "QUIT" key.

The settings for Altimeter, Geocaches, Marine and Fitness have not yet been mentioned. We do not want to delve deeply into their entirety. They are explained in the manual as well and also only apply to specific user groups, for which the available selections are already self-explanatory, anyway. In Chapter 3 / "Navigation," we'll take a closer look at the settings for Routing. Therefore, only a few selected points:

- **Altimeter**: Leave the Auto Calibration "On" and the Barometer Mode on "Variable Elevation," so that GPS64 always calibrates this when the unit starts and can determine the change in elevation, while you are moving.

- **Marine**: For use at sea, it is necessary to ensure diverse settings are enabled, so that, for example, a Marine Chart will be displayed exactly as required.

- **Fitness**: If you want to pair a Garmin heart rate belt, or a cadence sensor with your GPS64, the respective options must be turned "On" here.

"Bravo !!!" - You have gotten to know your device very well and simultaneously selected all the useful, real-world settings. Therefore, you can now leave the "Setup" menu with a press of the "QUIT" key and transition to the much more practical part.

The GPS64 offers a wealth of current and accumulated data for your tour progress. To find this numerical data during the tour, you open the "**Trip Computer**" view with "PAGE." Here, you'll find all possible numerical values that can be displayed according to your wishes. If there are too few data fields for you, you can also change the entire layout. Press the "MENU" key (2nd image from the left) and initially select "Change Dashboard." Navigate with the rocker key to the line "Small Data Fields" and confirm with "ENTER." With this, you get a display with 10 numerical values, like in the image on the far right.

Figure 2-14 Change the Trip Computer view via the "MENU" key

→ Other examples: With "Large Data Field," instead of the upper 4 small data fields, you can display 1 large data field. By selecting "Stopwatch," workout-oriented GPS64 users, for example, get the exact time on the display. This means that during the normal display of Trip Computer mode, you additionally have immediate access to the Stopwatch without first having to leave the "Trip Computer" display and the call up the "Stopwatch" display. Once you start the Stopwatch by using the "ENTER" key, you again have the possibility to divide times into laps. Pressing the "Lap" button, which is only displayed after the Stopwatch is started, gives you the ability to further subdivide sections of the tour in time and in distance, while the total elapsed time and distance continue to run. However, the stop times in detail, i.e. the

2–63

current stop time with distance and overall stopping time with distance, will then be found only in the "Stopwatch" view (accessed from the Main Menu or "PAGE" when placed in the Page Sequence). The times measured with a Stopwatch are not stored anywhere and are also not visible later on the PC. ←

Back to Trip Computer display: The values in the individual data fields can be changed by pressing the "MENU" key again, but this time choose "Change Data Fields." In the next display, a colored background appears on a data field. With the rocker switch, navigate to the field whose value you want to change and confirm the selection by pressing the "ENTER" key. Select a value from the list that appears and confirm it again by pressing the "ENTER" key, so that the selected value is accepted. If you are done with the entire selection of all data fields, you complete this step again with a keystroke on "QUIT" (of course), so that no box is highlighted. You'll find an overview of all available data field values, with explanations in the user manual.

Especially practical data fields are, e.g.:

- "Distance to Next" (with an active route, the distance to the next intersection is counted down - as long as you move in that direction - and you are fully prepared to curtail the conversation with your companion a bit, in order to concentrate on the turn);
- "Distance to Dest." (with an active route, or a track acting as a route);
- "ETA at Destination" (with an active route, or track);
- "Elevation";
- "Ascent - Total";
- "Vertical Speed" (rate of ascent, or descent over time);
- "Pointer" (an arrow showing the way to the target / direction of travel, during an active route);
- "Heart Rate."

And of course, the indispensable and well-known travel data we know from a bicycle speedometer, such as:

- Speed;
- Speed - Maximum;
- Speed – Moving Avg.;
- Trip Odometer;
- Time of Day.

In order for the values of each data field to display the correct value during the progression of the tour, the values have to be reset back to an initial value of zero, just before the start of the tour. To do this, go to the Trip Computer display, press the "MENU" key, then select the "Reset" option and then "Reset Trip Data."

Perhaps you already noticed it during the set up of the map display and intensively studied the "Geocaching" section. For everyone else, here again the clear note: You can also add some of the data fields from the Trip Computer display to the map view, in order to avoid permanently switching back and forth between the map and Trip Computer. To do this, press the "PAGE" key, until the "Map" item appears in the middle of the Page Sequence and then opens. Now press the "MENU" button and select "Setup Map." In the map setup menu that opens, select the "Data Fields" item and then press the "ENTER" key. Here, you can select between 0 to 4 data fields, or the Dashboard of, for example, the Stopwatch, Compass, Elevation Plot, Geocaching, etc. By selecting "Custom," you can specify whether the data fields are always displayed, or to be hidden during navigation, because the navigation text will still appear in the map, making the visible map area quite small.

With different types of uses

If you use your GPS64 for completely different activities, such as on the one hand, in the car in order to let yourself be guided to the starting point of the tour, and on the other hand on the bike, with which you then want to follow your prepared track. For this, you will need very different data field values and even device settings. Do to this and in order to relieve yourself from the requirement of constantly changing your settings, when you move from your car to your bike, it's better to make use of different <u>Profiles</u>. This means that all of the settings, which you have seen so far, such as

- enabling the topographic map and disabling unneeded road maps,
- the Page Sequence,
- the settings for route calculation,
- the display of the Trip Computer and its data field values,

can be chosen completely differently, after switching to another profile. You only need to call the Main Menu category "Profile Change" and either choose an already existing profile (2nd image from left), in which some preferences were previously set at the factory.

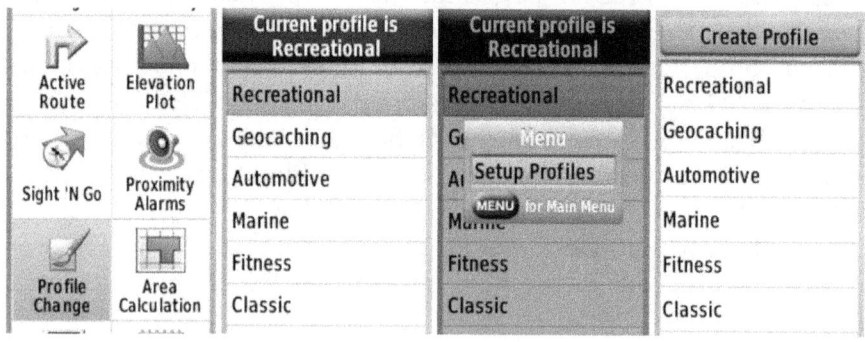

Figure 2-15 Use the GPS64 with a different profile / 3rd image from the left: via the "MENU" key to create a new profile

You can also create a copy of your currently applied profile, in order to transfer your hard-won settings, slightly modify them and not have to start over from scratch. For this you press the "MENU" button and select the only option available "Setup Profiles" (Figure 2-15, 3rd image from left). "Create Profile" now appears on the top of line (4th image from left). Using the rocker switch, jump up to this line, so that it is highlighted and confirm your selection by pressing the "ENTER" key. With this, your currently used Profile was copied and titled with the name "Profile 7." This can now be found at the bottom of the list. By selecting it and confirming with "ENTER," you can change the name.

If you would like to remove a Profile from your GPS64, you should select it in Setup > "Profiles." After pressing "ENTER," select the "Delete" option, which appears. The most recently added profile will try to fight back a bit, however, because it would like to be used first before it lets itself be deleted again.

In a nutshell, what remains to be said:

- In the main menu in the category **"Profile Change,"** you can quickly switch the GPS64 to a different profile with your especially elaborate settings, views and data fields, for use for the upcoming activity.

- In contrast, by pressing the "MENU" key in the Profile Change menu, you end up in the same place as you would from the Main Menu > Setup > **"Profile."** Here you can change the existence of the profiles. You can change the Profile names set at the factory, change the order, or remove and create other custom profiles.

Bluetooth – Utilization

All models of the new GPSMAP64 series with the model extension "s" and "st," feature a Bluetooth interface with which the coupling to a smart phone is possible. This allows you to benefit from the following additional features:
- See alerts from your smartphone on '64 display,
- Live Tracking: people you have notified can see your current GPS position and experience your tour in real time,
- GPS objects can be transmitted from the smart phone to the '64, or vice versa.

You will definitely need a Bluetooth and Internet enabled mobile phone with iOS, or Android operating system.

Receiving messages

In order to receive these on the display of the GPS64, select the GPSMAP 64s(t) Main Menu: Setup > Bluetooth > Connection "On" and "Messages: Show" (1st image from left). In the Bluetooth settings of your smartphone you must also allow the coupling of the '64. By now, your phone will have noticed the "knocking attempt" of your GPS device, and you should now find this device entry there. Tap on this and allow the coupling. Simultaneously, a code on the display of your GPS unit appears, which you have to type into your smart phone. Done. Both devices are now connected.

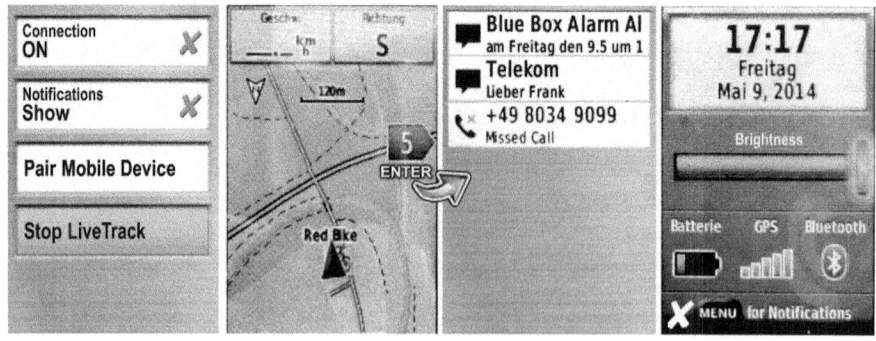

Figure 2-16   Set up Bluetooth and messaging

You can display incoming message notifications (Figure 2-16, 2nd image from left) by pressing the "ENTER" key (3rd image from left). You can again find messages, which were accidentally tapped away, on the Status page (briefly press the POWER button) with a press of the "MENU" key.

Live-Tracking

Live-Tracking can be used by ambitious bikers who want to use it during a race to send their actual position directly to the mobile phone of a supervisor, or team partner, so that they can follow the proceedings live, and know when the rider reaches a certain position (e.g. for personal provisions, rider change in team races, etc.).

Most users of a '64 model will certainly pursue other interests. There are other very good reasons for which you can use Live-Tracking. Namely, for example, for your safety in Alpine terrain. Before starting your tour, you can send an email to those at home, so that they know where you are, for example. In a power-saving manner, the smartphone sends the collected GPS data to the net every 60 seconds. If anything happens to you en route, the performance duration of your phone remains relatively long. After turning off, your last signal will still be able to show your last activities for another 24 hours.

For Live-Tracking you need the free app, "Garmin Connect™ Mobile," on your blue-tooth enabled phone. You can find all information about this application at:
www.garmin.com > Products > Apps.

If you have already installed this app on your smartphone, please open it and create a free user account, if not already done (Page 7 "Get started with"). In the Bluetooth settings of your phone, is your GPS device already paired, as just described under "Receiving messages?"

Open the menu by pressing the ☰ button of the app and select "**Devices**." Your device should be displayed here already, but it will not be released for use in Connect. You will probably be given the choice to "Add." Tap this word, which will activate this device for use with the functions of the app.

If the device is still not displayed, tap the "Devices" row and then the "+" sign (for adding) at the top right. Here, select your '64, which you will find in the device category "Hiking."

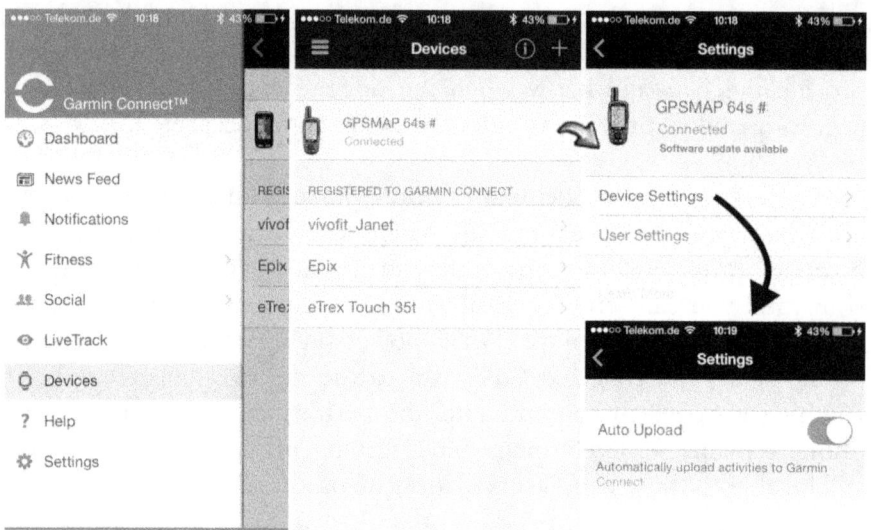

Figure 2-17
Garmin Connect ™ Mobile App: device set up

→ It is important to activate "Auto Upload" in the device settings. Only with this set, can observers see your current location, as well as the part of the journey already traveled. ←

Return to the menu again with the ☰ button in the app menu. Now choose "LiveTrack".
Enter a name for your activity and then tap in the "Invite Recipients" line if you want to send a note about your activity to selected individuals via email, or simply activate the corresponding social network, where your account activity is then seen.

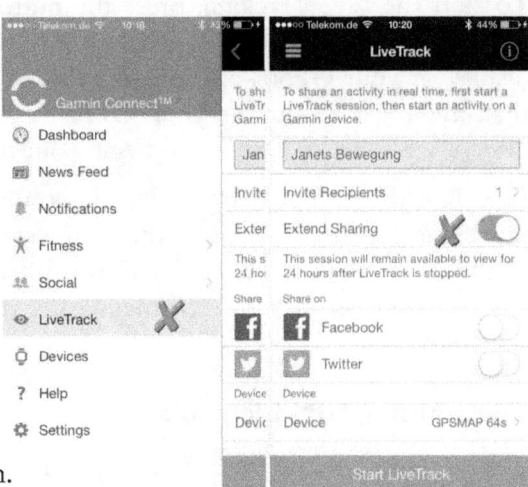

Figure 2-18 Start LiveTrack

→ Important: Also activate "Extend Sharing" here, so your activity can continue to be accessed 24 hours after completion. This also gives you the opportunity to look at your record of the tour, as seen through the eyes of your observers. ←

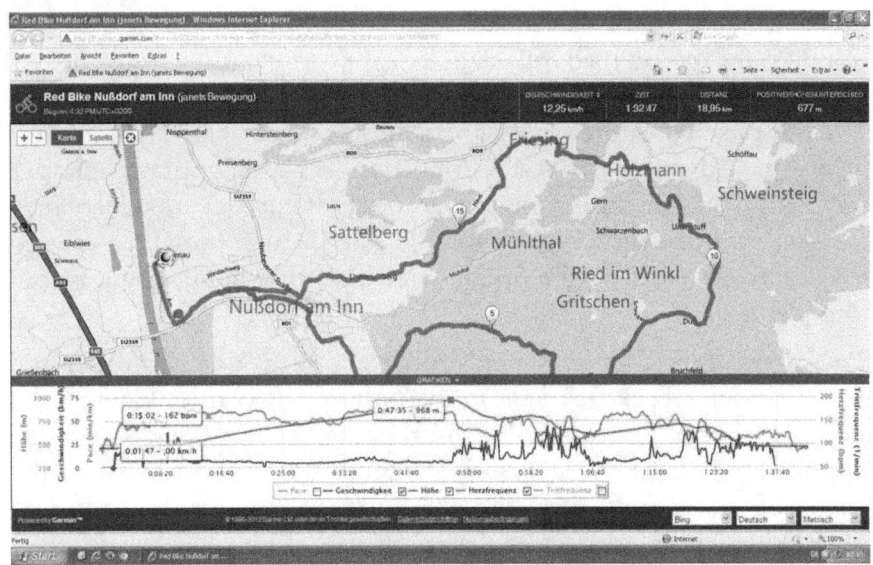

Figure 2-19 A Live activity displayed on the PC

2–71

To start the Live-Tracking, press the button on bottom of the screen of your smartphone. The LiveTrack note will then appear on the display of your GPS64. Confirm with the "ENTER" key. Make sure that the track record in your GPS64 is active (Setup > Tracks > "Record, Show On Map") and begin your tour.

Live-Tracking can be ended solely on your smartphone by selecting the "Live Track pause" button in the Connect Mobile app. A start, or stop of the track log in the GPS64 **will not** interrupt the Live activity. Only when track recording is turned off, will no recording data be transferred.

Transmitting GPS Elements

You can send, or receive, Waypoints, Tracks and Routes to/from the GPS64 with another free application, the "Garmin BaseCamp™ Mobile" app, which can be found at: www.garmin.com > Products > Apps.

This app is available only for smart phones and tablet PCs with iOS operating system 5.1 or later (for example, from iPhone 4S and iPad 3rd generation).

The most useful feature of this app is probably the ability to search for "**Garmin Adventures**" (GPS tour data with description, photos, waypoints, etc.). Because with your Internet-enabled mobile phone, you can spontaneously browse through the Garmin portal for offered tours, transfer them to the GPS64 and take off without any up-front tour planning.

To do this, open the BaseCamp Mobile app on your smartphone and select the small backpack icon in the bottom-left corner (Figure 2-20, 1st image left). To search for tour packages in the map, tap the magnifying glass in the upper right corner. Position the map in the region of your desired excursion area and wait for a few seconds until sufficient symbol characters have opened. Due to the different characters and colors of the small icons (2nd image from the left), you can already see whether it is related to hiking, climbing, skiing, cycling, etc.

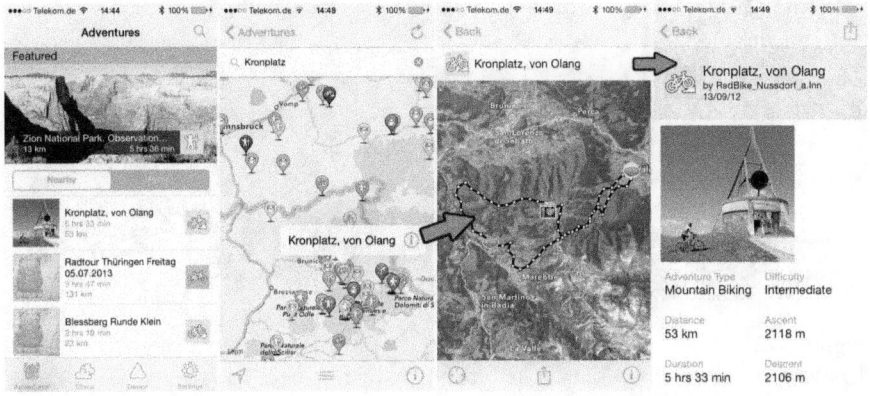

Figure 2-20 BaseCamp™ Mobile app: search for tours online

If you tap on a tour symbol, the tour name appears. If it already sounds full of promise and you want to learn more, tap the name line that appeared. The tour, with all the waypoints, will then open in a satellite image (3rd image from left). You can call up all details about the tour, as well as the description, by tapping the line above the satellite image. If it appears to be something you'll like, you can transfer all the contents of this tour package by clicking on the send button in the upper right corner of the detail page (4th image from left), or at the bottom of the satellite image page.

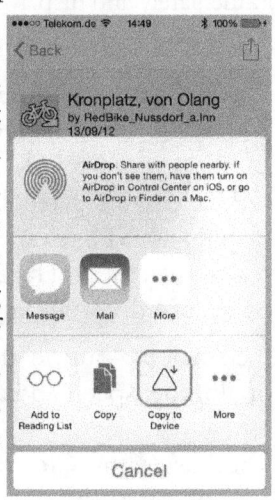

Figure 2-21 Smartphone icon "Copy to Device"

To check whether everything has arrived properly to the device, or if any GPS64 data can be seen on the smartphone, select the "**Device**" icon at the bottom of the screen of the BaseCamp Mobile app. All GPS data from your GPS device is immediately listed on the mobile phone display. (If nothing budges, you can also force a data transfer from the GPS64 Main Menu > Share Wirelessly > "BaseCamp Mobile".)

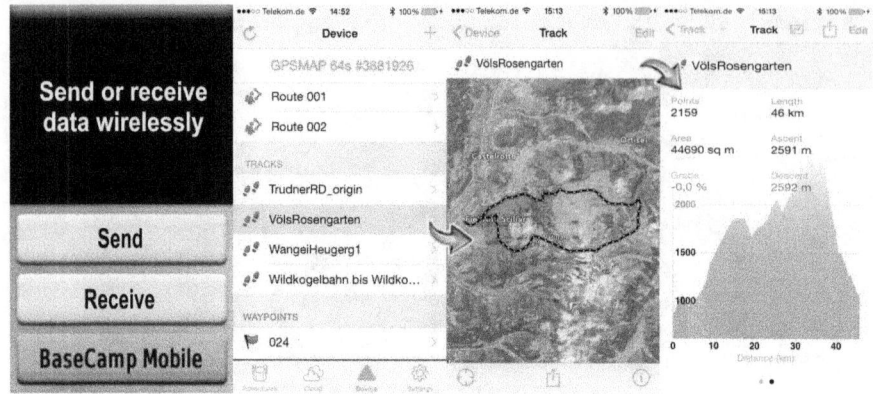

Figure 2-22
Image 1: Force GPS64 data transfer, normally not needed
Images 2-4: BaseCamp Mobile app: Data from the GPS device is loaded immediately and displayed on the "Device" page (photo 4 altered for graphic clarification)

Via the clockwise-pointing arrow ↻ in the upper left corner in the 2nd image from the left, you can retrieve the data from the GPS device anew, in other words, update the data list on your smartphone (if new GPS elements have been added, or edited in the GPS device). By tapping a GPS element in the list, the Satellite will open with the selected element. Here, a track is shown. Detailed information can also be accessed by tapping on the name line on the satellite image. With the small ⌦ button at the top of the screen, you get the profile display of altitude, speed and all sensor values.

The BaseCamp **Cloud** is another function of the BaseCamp™ Mobile app. Here, GPS elements that have been sent from the BaseCamp software on your home computer to the cloud, are now accessible via smartphone from the road and can be transferred to the GPS device.
In order to use this feature, you need to sign into a Garmin account. This is the account with which you also register your GPS device, and when you want to install and activate a Garmin map DVD on the PC. If you have not already done so, you can now also create the new account directly via your phone. Additionally, select the "Cloud" symbol at the bottom of the screen of your mobile phone in the BaseCamp Mobile app.

Figure 2-23
Use the BaseCamp cloud

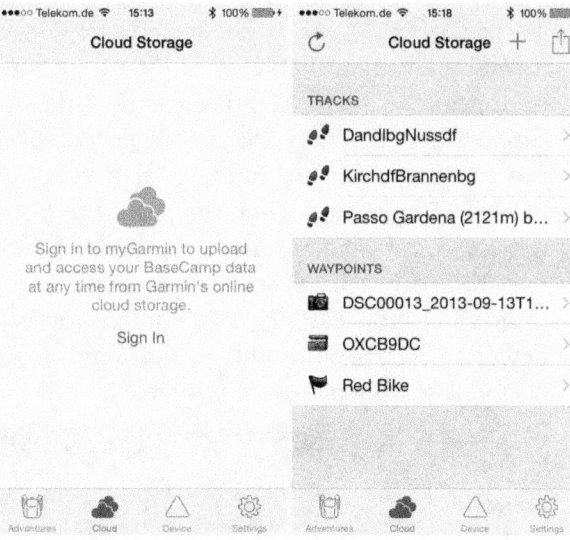

To now transfer a Track, a Route, or a Waypoint from the cloud storage displayed on the smartphone to the GPS64, you may have to do the following in the Main Menu of the GPS64:

Share Wirelessly > Receive > BaseCamp Mobile (However, it is also possible that the transfer works without this extra step.)

On the smartphone, start the data transfer again with the ⬆ send button and then select "Copy to Device."

The GPS data lying in cloud storage can be cleared by calling up the entry by tapping it from the list, then select "Edit" displayed at the top. Then tap on the trash can, which is shown at the bottom of the screen.

Pairing to a VIRB™ Action Camera

To use VIRB (Action Camera), or VIRB Elite (GPS Action Camera) made by Garmin, the GPSMAP 64s... models have a "VIRB Remote" option in the Main Menu, created at the factory. Numerous, diverse support systems exist for this resilient high definition camera. These can be placed in any number of hard to reach places on a bicycle, helmet, upper arm, backpack, or anywhere else. Due to this, the remote control is very useful in the GPS device. With it, you can start recording a movie, as well as shoot pictures directly from the GPS64, into which you can calmly position yourself. Because with the GPS device in hand - and with it, the remote control to your VIRB - you decide when the picture is triggered.

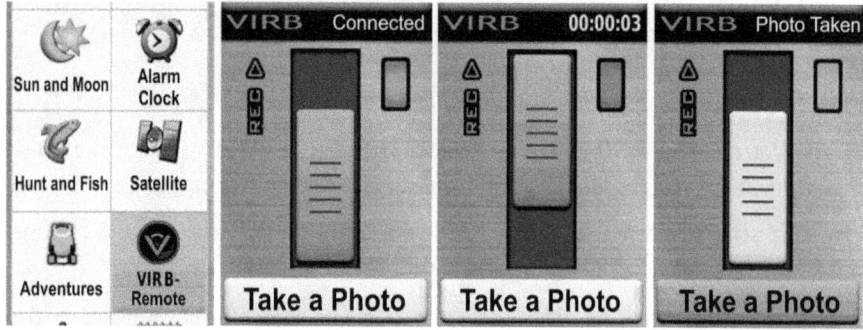

Figure 2-24
GPS64 Main Menu > VIRB Remote > REC, or Take Photo
Slider up: Record a movie
Button down: Take a photo

The VIRB is a rugged outdoor camera that combines image resolution and image stabilization in the forefront. Even when attached directly to the bicycle handlebars, the camera delivers great shots.

The most interesting point should then probably be the combination of imagery and GPS data. Mainly because the audience now has more of an ability to place themselves exactly into the situation on the basis of images showing road slope, elevation data, speed, exertion (based upon heart rate, or performance data) and any other information, when actually viewing the action.

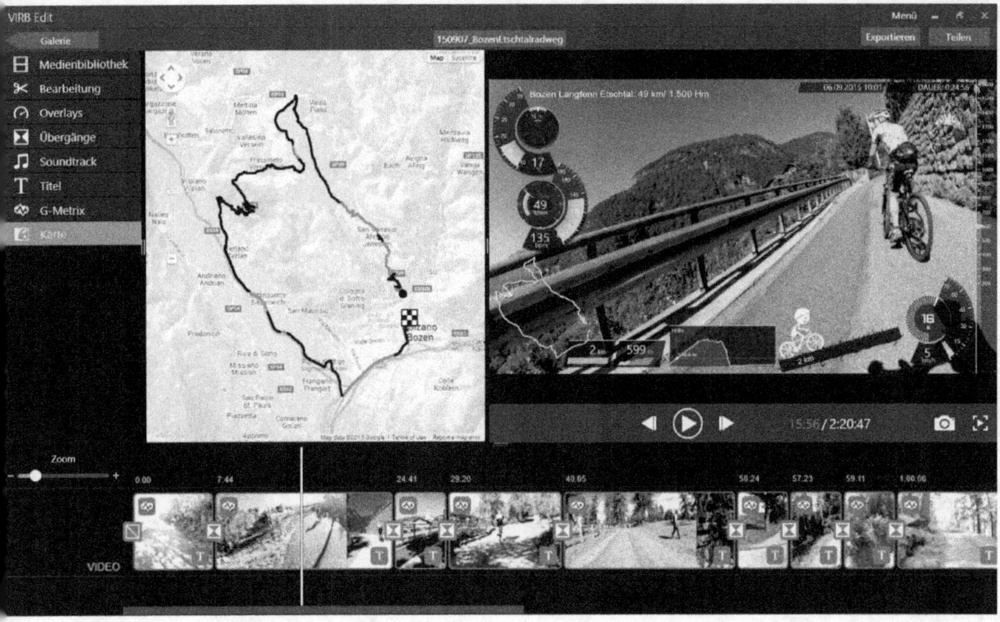

Figure 2-25 In the Garmin editing software, "VIRB Edit," you can cut the video clips, reassemble and select the display of GPS data

Figure 2-26 Final video from a VIRB™ GPS Action Camera

2–77

Chapter 3 - Navigation

As already mentioned in Chapter 1 ("Routes and Tracks"), you have several options for navigating to a specific destination with the GPSMAP 64 models:

- Either you select the mode of transportation and destination in the device. The device subsequently calculates the way, i.e. <u>Route</u>: The device sounds a beep at every fork in the road and turning instruction. If you move away from the route, due to the deviation, this will be immediately noticed and the route will be corrected/recalculated. You work with a route, when the destination is selected spontaneously while underway, or you don't want to worry about the way you will get to the destination.

or

- You have downloaded a <u>Track</u> from the Internet, or created one on the PC itself, uploaded it to a GPS device and this track is shown as a line on the display: The unit is silent and you have to make sure where you are and where the track line runs. When moving away from the track, the track line wanders from the display field of view, but remains unchanged and is never recalculated. You work with a track, when you have a clear idea about the itinerary, because you have heard/read of this dream tour, for example.

The advantage of **Track navigation** is that you can deliberately move along the planned paths to reach the goal. The track is shown as unchanged in the GPS64 and never recalculated. You can immediately use a track in the unit, without the need of any maps.

The advantage of **Route navigation** is that you do not have to make any preparations on the PC, before the tour. The destination is selected in the device at the beginning of the tour. The way to get there is calculated and created by the device software itself.

Let's start with the variant that the advertisements so enticingly describe with "Turn On & Go," in other words

Route navigation

= automatically leading the way (Route) to the destination. The road is secondary. The shortest and best way is calculated by the GPS64. It is, more or less, at the mercy of technology.

We have learned: A route is suitable for navigating in one direction and primarily serves spontaneous target selection in the device (on the go). In order for the device, or rather the software of the device, to calculate a route to the selected destination point, it requires <u>routable maps</u>.

As described in more detail in Chapter 1 ("different types of maps"), there are a number of different maps and different ways to install them on the device. The exact procedures are described in detail in the next chapter 4 ("Map installation").

So, we can assume that we have "fed" the GPS64 with maps, either placed a pre-programmed microSD card in the card holder, or have even uploaded parts of maps from the PC via the Garmin mapping software.

➔ If you have different types of maps (i.e. road and topo maps) with the same coverage area on the microSD card, or the device storage, you must also remember that the route will be calculated based on the top-most, visible map. Topographic maps are arranged on top of road maps in GPS and must be disabled if the GPS device is to calculate the route on the road when travelling by car. In order to not overlook this, it is advisable to only activate the map, which is required for the current activity. For this reason, use of different profiles for each activity is a great advantage. ←

Another point for the correct calculation of a route is the mode of transportation for which the route is to be calculated. This is because the GPS64, of course, calculates routes for touring in other ways as routes for a mountain bike, car, motorcycle, hiker, etc. To do this, go to Main Menu > Setup > "Routing." Here in the **routing settings**, you have the possibility to adapt the route calculation to your activity:

- The Activity: Mountain biking, mountaineering, etc. (Direct Routing is also possible.)

- The Calculation Method: You can choose between the version of the shorter distance, shorter time, or a gentler ascent. When you select "Prompted," with the start of each navigation, you will be asked which type to use.

- Lock On Road: The position arrow and recording will be fixed on the road center. In the field, the track record can quickly jump to an adjacent path. Therefore, setting this to "Yes" is to be treated with caution.

- Avoidance Setup: Depending on the selected activity, your preferred avoidance requirements can be precisely set based on the specific situation. For example, by no means do you want end up on unpaved roads with a car, or bike. The "Carpool Lanes" item is unlikely to be of importance to those in Europe. This road provision is principally encountered in America, meaning that this lane must not be used by cars with a single occupant (driver only).

- Off Route Recalculation: With "Automatic," the GPS64 calculates your route once again, as soon as it detects that you have moved from the proposed route. With "Off," the route remains in its original state. With "Prompted," you have the option in each individual situation to determine whether the route will be left as is, or recalculated.

Tour start - Route

The device is now adjusted to your movement type and by using the "FIND" key, you can choose your destination from the extensive selection objects, or POIs, and start navigating there with "Go."
If there is a road map, or the Topo Germany V6 installed in the device, entering the exact address is possible. But even waypoints you've stored yourself, tracks, routes, photos, coordinates, recently viewed items, restaurants, mountain huts, etc., to which to navigate, can also be called up in the destination entry subgroups. Which POIs are available for selection also depends on the type of map installed.

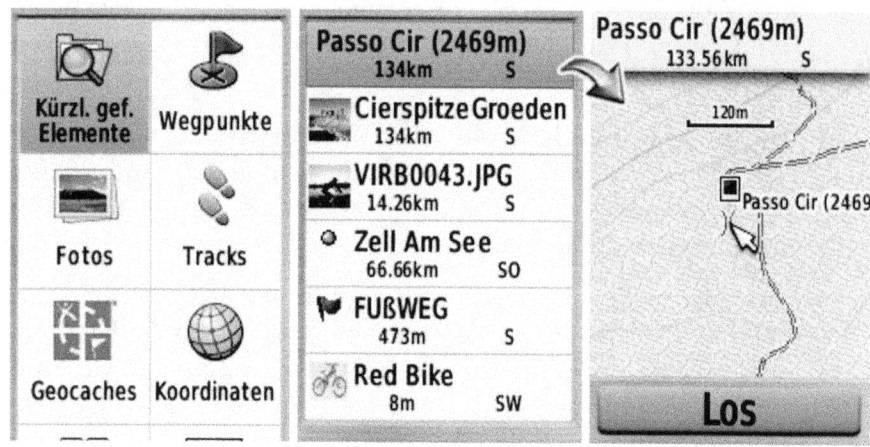

Figure 3-1
e.g. Use "Recent Finds" to select the target destination

The search in each destination entry subgroup always originates from your current location and displays the results in order of increasing distance. Example: You are currently on a long bike tour and notice a slowly creeping feeling of hunger. Due to the length of the tour, it would a good opportunity to take a break in roughly 20 km (12.5 mi). The following paragraph shows you how to find the rest stop locations, which lie ahead:

"FIND" key > now immediately press the "MENU" key: Select "Search Near," and in the next step select the "A Map Point" option in order to be able to move the pointer to the place in the map where you want to search for some "food and drinks." Then confirm with "Use" and from the reappearing destination menu POI items, choose the category "Food and Drink" > "All Categories." The search begins, but this time **does not** start from your current location.

In a similar fashion, you can also do this directly from the normal map view. With the rocker key in display mode, aim the pointer at a particular region, then press the "FIND" key and now select the desired POI category.

➔ Tip: In order to quickly select a destination while underway, it's more convenient when you create waypoints out of the candidate targets at home on the PC (stored in a GPX file) and transmit them to the GPS64. This way, you can promptly go to "Waypoints" in the destination menu and start your navigation to there. ←

The following points should always be followed, when starting a navigation:

- turn on the GPS unit under an open sky, in order to work with the best GPS reception,
- the routing settings have been chosen correctly and the correct map is activated,
- at tour start the data field values in the trip computer view have been reset to zero in order to be able to show the correct travel data and
- recording is active, if you wish to have it.

Apart from that, now simply follow the magenta colored route line on the map and the instructions at the top of the screen. By calling the "Active Route" view (in the Main Menu, or by adding it to in the Page Sequence), you can display the full list of all upcoming turns, call them individually and receive information about the turning situation in advance. Those who like can also navigate based on this list, because at least one message, with the graphical turning instruction, is also automatically displayed in this view, before each turn.

Figure 3-2 Call the "Active Route" page, in order to show the turn preview list

Changing the active route

If you want to change the method of calculation during an active route, this can be easily done via the "FIND" key. This is because the "Change Routing Activity" choices, to which you can switch, are here. Examples are Direct Routing, Pedestrian Walking, Hiking etc. If, during an active route, you decide that you want to first hit another point in between, press the "FIND" key and select the "Find Another."

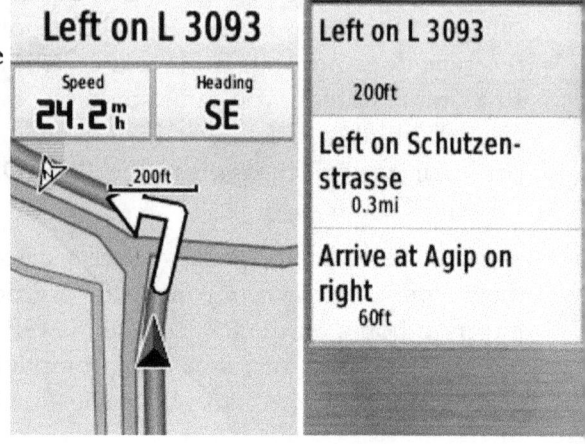

If you have reached the end of the navigated tour and the target was not detected automatically, you can stop the navigation by using the "FIND" key and selecting "Stop Navigation." You can tell whether you have an active navigation by pressing the "FIND" key. If you do not receive the list of categories, other than those to which to navigate from the destination categories menu, then you do not have an active route.

Prepare a route in the device

With the "Route Planner" you can create a route in the GPS64 whose starting point can be any arbitrary position (Main Menu > Route Planner). By activating the top line, "Create Route," the usual menu will open and you will have the opportunity to enter a start point, via points and end point of the route, either through the selection of stored waypoints, or those in the POI collection, or by searching in the map. If you have confirmed all your interim points with the "Use" button, leave the "Create Route" menu with a press of the "QUIT" key. You return to the "Route Planner" menu. Immediately the newly created route is stored with a number in the list below the "Create Route" line. If you want to change something later, you can select this new route again and open its properties by pressing the "ENTER" key. In the menu which appears, the following can be done:

- "<u>Edit Route</u>," for example, add another via point, or change the order of the points: By clicking on the existing point, another submenu opens, where you move the order of the point up, or down, or delete the point. With "Review," you can simply see the point on the map and then go back with "QUIT.";

- view the route in the "map" and start with "Go";

- show the "Elevation Plot";

- change the name of the route;

- <u>Reverse Route</u> and

- "Delete Route".

In the "Routes" selection within the destination selection categories menu after pressing the "Find" key, you can call up the route you have just created in the Route Planner for navigation and start with "Go."

➔ If you are not yet at the planned starting point of the prepared route and start it with "Go" anyway, the path to the normal starting point is simply prefixed and forms an entire route with the planned route. The situation is different if you are already in the middle of the planned route and only at that point start to navigate with "Go." Then, misunderstandings arise and your GPS64 sends you back to the initial starting point of the route, in order to be able to follow the route from there, with all its interim points in the correct order. So, you should probably only use a prepared route, if you begin at its starting point, or at least haven't overshot it. ⬅

Track navigation

= the "trail of bread crumbs" (Track) that has been recorded with a GPS device and to which you would like to go back to and do another time. It may also be a tour prepared, or plotted on a PC. A track cannot be changed in the device. Navigating via a track is particularly suitable when the starting and destination points are identical, which is standard practice in a bike ride. The file must be in GPX format and can be found in the GPS64 under Main Menu > "Track Manager."

When navigating via a track, you must ensure that your position arrow stays on the track line. If you move away from the line, the line moves off of the display, because of the position arrow is always at the bottom center of the screen. Under no circumstances will the track be updated with a new calculation and always remains just as you created it on the PC. An installed map is not necessary for this purpose, however one would help immensely when needing to decide which way to go at a fork in the path.

Displaying multiple tracks in different colors (black and red, here) is the best way to show the planned path, as well as any optional paths in case of tour terminations, on the spot decisions, or shortcuts. The turquoise line marks the actual path traveled.

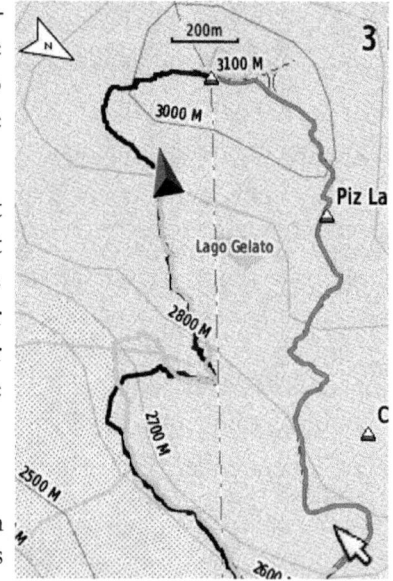

Figure 3-3 Map view with different colored tracks

In the '64 you activate the visibility of a track in the Track Manager > Select the track > Row 3: "Show On Map" and confirm.

It's probably one of the most common reasons why you would ever a buy GPS device: there are countless tours, which have already been recorded by others, available for free download (in some cases, paid) on the Internet. Included is an enticing itinerary and then: "Yes! - That's exactly the right one for me. That's a tour I absolutely have to do."

Without extensive map studies and path exploration, you load the tour into the GPS64 and away you go into an area where you've never been and yet still know where to go. That's pure recreation enjoyment.

So you've decided on a tour from an Internet portal. Even though the download itself is free, it may be that you have to at least register in order to reach the section to download the GPS files.

With professional tour portals like www.gpsies.com, among other things you have the possibility to send the track directly to the device. Your PC requires a small tool (plug-in) in order to be able to make the communication link between the portal and GPS device. This "plug-in" will be automatically installed when you allow the displayed warning. If the download actions do not run properly using Windows Internet Explorer, you may be better off using the free "Google Chrome" browser.

It makes more sense and it is always safer to first store all downloads from the Internet on your own computer. In this way, you keep a better overview and you should view the downloaded track in a GPS map on the PC, before sending it to the device, anyway. At this point, you can for example, delete any unwanted parts of the path, or perhaps supplement it with some small improvement, which you would prefer to include in your travel.

If the track is finally as it should be, save it again as a GPX file on your PC and copy it to the GPS device storage as follows:

Send data from a PC to the GPS64, without GPS software

Connect the switched-off GPS64 into the PC via a USB cable and wait until the external drive is automatically recognized.

A Windows Explorer window should automatically open with the "Garmin GPSMAP 64… (E:)" drive detected, or the next available drive letter. If you have the impression that your PC is not responding, you can open the drive manually via "Start" > "My Computer," or by double-clicking "My Computer" on the desktop.

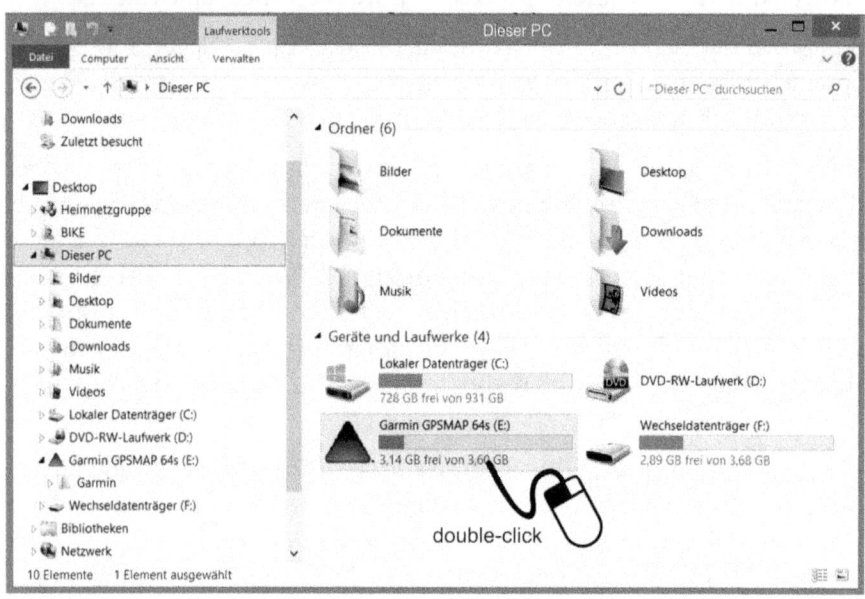

Figure 3-4 Access the GPS device memory and microSD card memory in Windows Explorer

The GPS64 device storage is indicated by the drive name "Garmin GPSMAP 64… (…)". If there is a pre-programmed Garmin microSD card in the device, another "GARMIN" drive is displayed. If another recordable microSD card is in the device, its card name, or "Removable Disk" will be displayed.

Now with a double-click, please open the device storage on the drive identified as "Garmin GPSMAP 64 (…)." Here, you can also see a "Garmin" folder with subfolders, such as "BirdsEye," "CustomMaps," "Filters," "GPS," "GPX" etc.

➜ **Caution - protect the data files:** (see Chapter 4 / "Create a backup copy of the GPS device storage"). Do not delete any files, if you have any doubt about their content. ⬅

Back to the sending of the track:

In the open window of the My Computer Windows Explorer, the GPX track file can now be easily copied with the mouse via the drag-and-drop copying process into the device storage of the GPS device.

Procedure: The easiest way is if you display the directory structure of the PC on the left side of the window. To do this, from the list of folders on the left side, click on the name of the folder, in which you have saved the saved track that you, for example, downloaded from the Internet to your PC's hard drive. The entire contents of that folder then appear in the right-hand pane of the window.

On the right-hand side, you should therefore see the saved track file that contains the track you want to send to the device. Now in the left-hand folder list, please scroll to the "Garmin GPSMAP 64 ..." drive (device storage), and with the left mouse button click on the triangle in front of the device name and then again on the triangle in front of

Figure 3-5 Copy the track as a GPX file

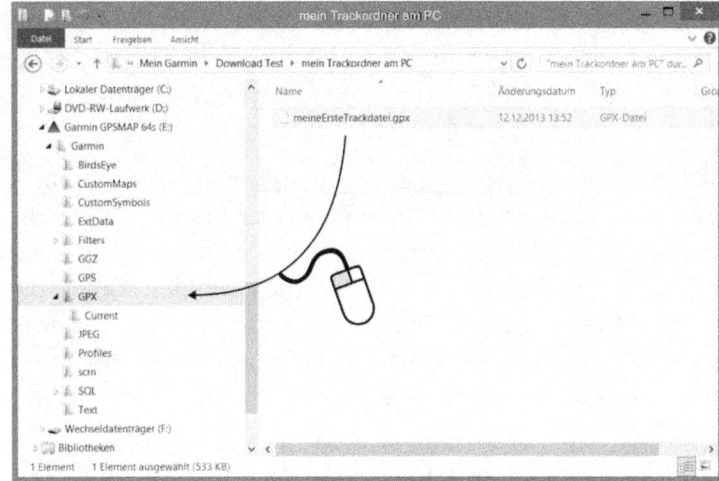

the "Garmin" folder until you can see the GPX folder in the left-hand list.

Now, in the right-hand windowpane with the left mouse button, click and drag the track (the GPX file) from the right window precisely to the to the GPX folder in the list on the left until it is highlighted in blue, and exactly at this point, release the mouse button. (You can also achieve the same results with the copy and paste function from the context menu, when clicking on the file with the right mouse button.) Done!

For verification, click on the "GPX" folder name in the list on the left. Its folder contents are displayed in the right-hand pane of the window, and you should see the copied GPX file there.

In the same manner, GPX files can also be stored on an empty microSD card that you have placed in the GPS64. For this purpose, however, the still empty memory card requires the same folder structure as the GPS64 device storage. The exact procedure is described accurately in "microSD card set up" in Chapter 4.

→ The name of the GPX file has nothing to do with the name of the track that appears in the GPS64 Track Manager. In the Track Manager, you will find the name of the track that was given when creating, or editing it in the GPS software on the PC. There can also be several objects (e.g. tracks and waypoints) stored in a single GPX file and copied to the GPSMAP. ←

The available storage space for tracks in the models of the GPSMAP 64 series is not limited to any particular number. Since the '64 internal storage, as well as the card storage can be accessed in the same manner as any external drive on the PC, they could really be inundated with tracks. However, in order to preserve an overview, or not to eventually have any loss in the start-up speed of the GPS64, it is of course advantageous to always only keep in storage, what is really still needed.

Now remove the hardware safely from your computer by using the Eject function in Windows (the green arrow in the Taskbar at the bottom right) and remove the GPS device from the USB cable.

Transfer via the ANT+ interface

Another way to get a track onto the GPS64, is via wireless transmission from device to device. For example, if your friend already has the tour available in his, or her ANT+ compatible device, you can bring both devices close together (<10m/30ft) and navigate to the "Share Wirelessly" category in the Main Menu. Your counterpart chooses "Send" in the window which appears, and of course you choose "Receive," in this case. At this point, you have to wait, as your acquaintance still has another step where he, or she has to choose what to send. Because it doesn't have to only be tracks, but also routes, waypoints and geocaches can be transmitted. When the process is started, you both see the progress of the transfer in the device display. After the sending process is complete, the information and processing options for the transmitted track can be found in the Main Menu in the Track Manager, waypoints in the Waypoint Manager, of course and routes in the Route Planner, etc.

The track is now in the GPS64 device's memory, but is not shown on the display in the map view.

Switch on Track visibility

What now confuses many and without intention is that they assume they now have to call the track with "Go" in the unit for navigation. No! You don't have to do this at all.

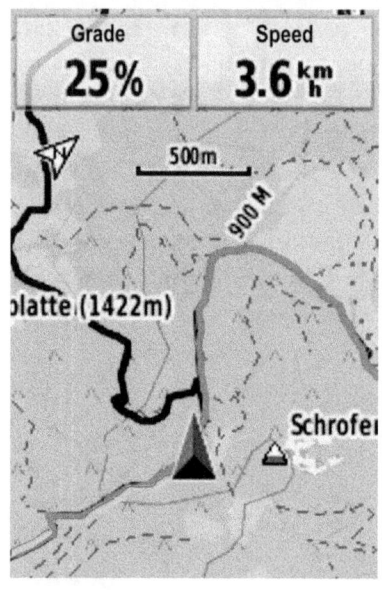

The only thing that is always important: we tell the GPS64, that it should display the track in memory, as a line in a color of your choice on the map. This must be carried out only once after transferring from PC to the GPS device. The next time the device is turned on, this "display obligation" for the track remains in place. If you have worked out additional, optional routes (because it is not yet certain whether the tour can be finished, or you have to eventually take shortcuts), choose another line color for them than that of the main track, so the optional route branches can be immediately recognized on the display.

Figure 3-6
Main track = red line
Optional track = black line
Actual recording = turquoise

Turning on visibility of a track line is best done immediately after working on it on the PC, and at the same time, in order to verify that it was correctly transferred to the GPS64.

Procedure on the GPS64: Main Menu > Track Manager > click the desired track, click the 3rd line in the track properties menu that appears:

- "**Show On Map**" – With this, you turn on the visibility of the track line in the map. The row text then changes to "Hide on Map," with which you will therefore be able to switch the now visible track, back to invisible again on the map.

Figure 3-7 Track Manager / List of individual track characteristics

By clicking on the row that immediately follows:

- "**Set Color**" – you determine in which color this track line appears on the map. Dark colors are usually best. (The color selected, when creating tracks in BaseCamp on the PC, is transferred and already present.)

That's all. From the Track Manager, go directly back to the Main Menu with a press of the "PAGE" key. With "PAGE" again, also open the normal map, where you should now see your line, once you are in the area. The settings for the track remain even after restarting the device.

3–93

Tour start - Track

Now you would just turn on the GPS device at the starting point of your planned tour and then immediately see the line of your tour course on the display. You wouldn't need to start any navigation. The device tacitly displays your planned tour (line) in the color of your choice, refrains from any route calculation, but also doesn't give any turn-by-turn hints.

In order to display the forthcoming navigation data on the basis of a track, e.g. how far you still have to travel, or to be able to see the upcoming profile colored in blue in the Elevation Plot, you need to start the track with "Go" (destination selection > Tracks) - theoretically like Routing. Basically, however, nothing happens other than the desired color of the track being replaced by the magenta-colored navigation line. Your optional tracks additionally remain visible in the color of your choice. Nevertheless, no turn-by-turn instructions appear, but you now enjoy the following benefits:

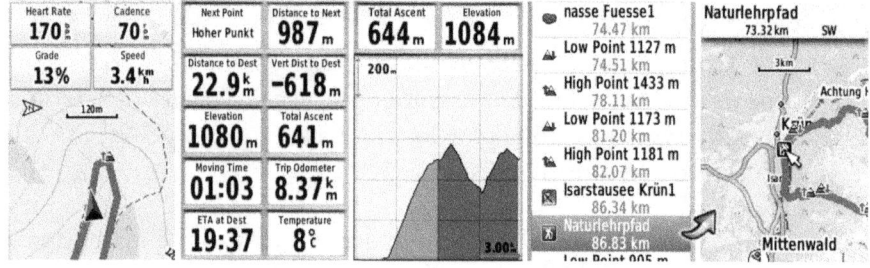

Figure 3-8 Start a Track with "Go"
Map with "High | Trip Computer | Elevation Plot | Active Route | Low point
Point" indicator | with ETA at | view | view | preview
| Destination | | |

In addition to the preview in the Elevation Plot, you now benefit from the fact that in the normal map view, the lowest and highest points of ascent and descent appear on the track. You can use the "Active Route" view to view the list of upcoming high and low points. This list will also show any waypoints detected on the track with their distance and arrival time. In the list, you can select any arbitrary point, in order to see it in the map (at the top edge of the map you will receive the distance and direction, as the crow flies). In the Trip Computer view, it now makes sense to display data field values such as "Distance to Dest.," "ETA at Destination" and "Distance to Next," because after starting the track with "Go," your current position will now be continuously calculated by the GPS64.

Track with turn-by-turn instructions

Still, there is a trick that exists to make your GPS64 pay attention once a turn appears during track navigation. For this you must however create waypoints at all turning points. That would be possible on the device, but is much better to do in the map software on your PC. However, it remains an additional effort, which only a few would really take into consideration. Test it out.

The fundamentals of drawing in "BaseCamp" comes in Chapter 4. Now, only some theory:
Display your track on the PC in the map software, and put a waypoint at every turn. Add some brief information such as, "left," "right," "water," etc., according to your wishes. At the end, save all these turn-waypoints to a GPX file. Now, use the Garmin POI Loader from the site: www.garmin.com > Support > Software > Show Supporting Software > Mapping Tools. With this little tool, which is installed on the PC, you can convert waypoint files into POIs, and also at the same time in this process, induce the alarm function. In the unit, you also need to activate the alarm for each individual waypoint (Main Menu > Proximity Alarms > "Create Alarm").

Launch the POI Loader on the PC and select the location for the POIs "Custom Folder," because we want to first save everything on the PC. The next step asks you for the location where the new, converted POI files are to be stored. In the next window, specify where the to-be-converted waypoints are located. Then select the distance units used to measure the distance from the point where the alarm should be triggered.

Select the "Manual" conversion mode so that you can enter the proximity distance itself in the next step (pictured right). Think about how far in advance you want to be alerted to a turn. Done.

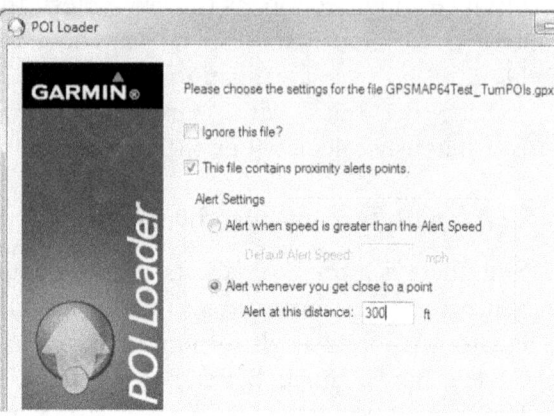

Figure 3-9 Convert Waypoints to POIs with Alarm function

Plug your GPS64 into the PC via USB. With the desktop explorer, create a "POI" folder within the "Garmin" folder of the device storage (see Chapter 4 / "Device storage: System and folder structure") and copy the newly converted "POI.gpi" file into it. You can rename this file as desired, e.g. Test.gpi. "GPI" must remain as the file type.

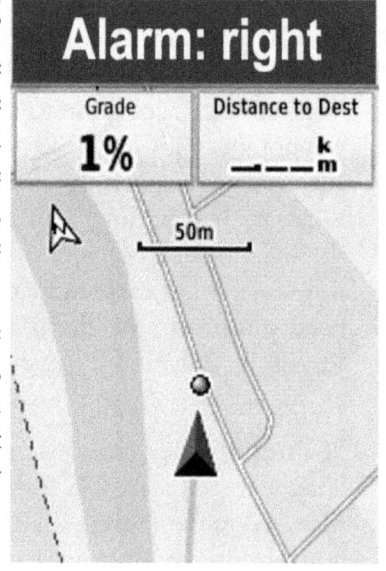

The alarms are immediately activated in the GPS64 and make their presence known, as soon as you enter the defined alarm area and the following are turned on: Set Up > Tones > Tones **and** Proximity Alarms.

Figure 3-10 Proximity alarm

However, when you travel along the track in the reverse direction, the turn POIs remain unchanged and would constantly report the wrong direction.

Sight 'N Go

Anyone who is moving around the out-of-doors, where perhaps not even a trail is charted on the map, or there is not any map at all of the area present in the GPS64, can well be served by using the "Sight 'N Go" function. It allows you to orient yourself on a point visible in the distance and navigate to it, based on the compass needle and straight direction line. This feature is also a popular venue in the marine sector to order to circumnavigate shallows, or other obstacles.

If you can easily see the destination from the position of your current location, but are not able to travel straight to it, you can use this function to the hold the target in the GPS64. In the map view, the original course (from the starting point to the destination) appears as a thick, magenta direction line. If you cannot follow this this ideal line, but first of all need to perhaps walk around a mountain, or descend into a valley, an additional thinner magenta line will be displayed. This is the bearing line that always shows as a straight line from your current position to the target.

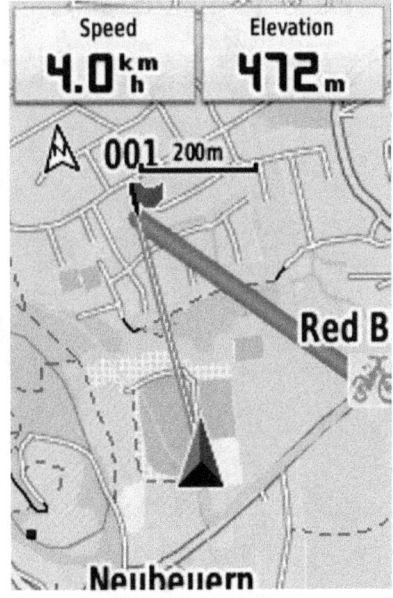

Figure 3-11 Course and bearing line in the Map view, going from RedBike to the targeted point "001"

Figure 3-12 Course and bearing line in the Compass view

Procedure: In order to aim at a target, open the "Sight 'N Go" application in the Main Menu. Hold the device in the direction of your destination and select "Lock Direction." In the window that appears, confirm "Project Waypoint" selection, choose the distance units in which you want to enter the estimated distance, and type in the numbers in the next step. Finally, select "Save," in order for the navigation to immediately start. As you desire, you can now navigate to the objective with the map, or compass view.

With the "Sight 'N Go" function, use the following compass setting: Setup Heading > Go To Line (Pointer) > "Course (CDI)." With this setting in the Compass view, you will see your original course direction represented by the arrowhead and its end on the one hand. On the other hand will see your current deviation from the course, represented by the varying center part of the arrow. This will separate itself proportionally to the distance from which you deviate from the course of the arrow alignment.

The "Sight 'N Go" function is also good to use to find out the names of the mountain peaks in the surrounding panorama, while on tour. To do this, you must simply be able to estimate distances accurately and can then look at the map, to see which mountain peaks are located at the targeted endpoint.

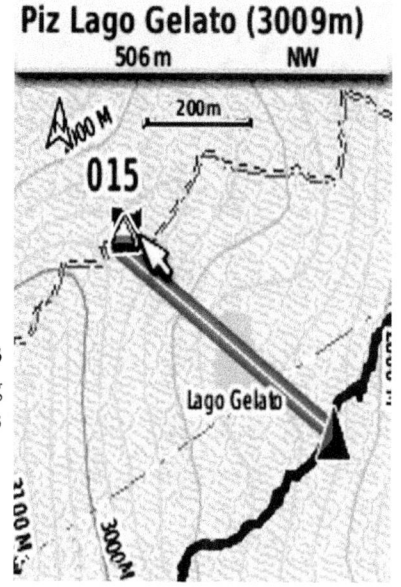

Figure 3-13
Aim for the surrounding peaks, to find their names

3–98

TracBack

Even if you have not called up a navigation in your GPS64 while on tour, have also not chosen the traversed track line to be displayed, but have only activated the current track log, the GPS device can guide you the same way in reverse (TracBack).

Main Menu > Track Manager > Current Track "View Map" shows all the GPS64 records that are still in the current track memory, as a reverse-traceable line and the "TracBack" button on, with which the reverse route planning can now be started.

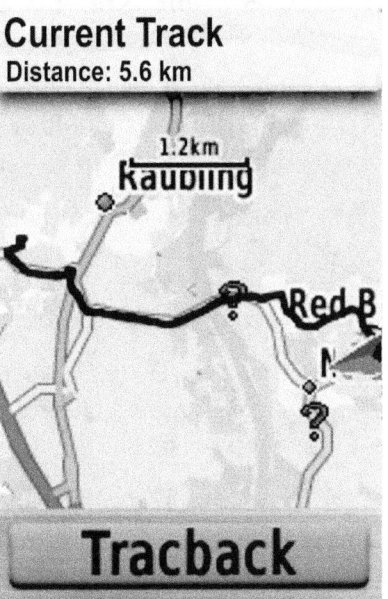

Figure 3-14 Navigate back to the beginning of the current track

This function is often initially ridiculed, but could be of lifesaving importance in an unknown holiday region, in desert, or glacier landscapes, where there is no path that can be seen, or in case of sudden change in weather in the high mountains.

So, nevertheless, keep it in the back of your mind!!

Tour start/Tour end - Steps in the Device

Without fail, think about the following three steps at the beginning of each tour:

- Display the prepared track, or start the selected track, route, or target with "Go" (in the case of the last two options, first adjust the type of movement in the routing settings),

- Reset the Trip Computer, best done in the Trip Computer view with "MENU" > Reset > "Clear Track and trip Data: Reset Both,"

- Turn on track recording.

After the tour:

- Turn off track recording,

- if the goal was not recognized, or not reached, end any navigations which were potentially launched ("FIND" key > "Stop Navigation")

- Save the Track by hand, in order to separate the track from subsequent recordings (in the Main Menu > Track Manager > Current Track > "Save Track"), then empty the current track log to avoid double memory usage.

Save the Track record

Actually, you do not have to worry about saving your tour recording, because you have chosen automatic storage in Setup > Tracks > Auto Archive, so the GPS64 performs a save either daily, weekly, or once the current track memory reaches 10,000 track points. If you are on the road for several days, you most certainly have no desire to tend to the care of the device. It is probably the case that you are the type of person who does post-processing work on the PC to cut out the required track parts from the full recording.

By manually saving, you can completely avoid post-processing after a completed tour. By manually storing, the track record is saved in another storage location in GPX format, namely in the GPX folder in your device storage, which can have an unlimited number of track points. You will find this separately saved track in the device beneath the "Archived Tracks" line in the Track Manager.

...and also, after connecting to the PC via USB, you'll find a GPX file in your GPS64-device's GPX folder with the date, or the name that you typed when saving.

If you did not empty the current track memory immediately prior to departure, but only zeroed out the data in the Trip Computer view with the "MENU" key > Reset > "Reset Trip Data," there are still some other track parts in the current track memory. Therefore, you have the possibility to fish out and save only a specific piece of the currently saved Track via Track Manager > Current Track > "Save Portion."

After manually storing, the track record is shown in your device in two ways:

- first in the "Current Track memory"
- and second in the device storage with the name as you personally just saved it.

Therefore, after saving, you are prompted as to whether you want to "Clear the current track." This removes all previous records from the "current track memory." In doing this, you will no longer have these double recordings in the device and will have already set them aside in a neat and recognizable fashion for transferring to the PC.

The "Archived Tracks" folder is a storage area which has 200 memory locations available, each with 10,000 track points. This space is allocated to the automatic swapping of the GPS64. If the current track memory is full, or the track Recording Interval is configured so that tracks are to be archived daily, or weekly, these track records are automatically moved from the current track memory to this storage archive.

Basically, you have three different storage locations in the device memory, so to speak: Current memory, archived storage and "free" available GPX storage (folder).

If you are on the road for several days, it is advisable to separately save the current record at least every two days and to empty the "current track memory." This is because you can expect to record 2,500-4,000 track points per day (normal track Recording Interval).

If manually saving while on the go is way too stressful, you can also keep a good overview in the GPS64, with the "Daily" setting in Auto Archive. In this way, a single file is archived every day and with 200 archival storage locations, there is no actual reason to be miserly.

But of course you can also leave everything in the raw state, trust the "When Full" Auto Archive selection and perform a "clean up" of the recordings later after uploading them to the PC in BaseCamp, or other suitable GPS software.

➔ With the models of the new GPSMAP 64 family, you can save 200 track recordings in the archive, with up to 10,000 track points each. In addition, you can load more tracks, routes, waypoints and geocaches into the GPX folder, as well as other GPS objects into their specific folders of the device and microSD card storage. ←

You can remove tracks from the GPS64 by going to the Track Manager, pressing the "MENU" key and selecting "Delete All Saved." All tracks in the GPS64 that were manually stored are removed. Or by calling a specific track and selecting "Delete," only the respective track is erased from the GPS64.

"Congratulations" - the introduction to the unit is successfully completed! You should now know:

- how to deal with basic GPS terminology;
- what types of GPS navigation controls your GPS64 has and how to use them;
- which device settings are especially interesting, or important for you and where to find them;
- how can you load tracks into your GPS device and show them on the display;
- how to start, or prepare routes in the device

and

- how to manage GPS recordings in device.

If there is something from the above list about which you know nothing, start again from the beginning. No, just kidding! Take advantage of the table of contents at the beginning of the book in order to find the topic, that may have become somewhat unclear in the interim.

Chapter 4 – Working on a PC

A mountain bike tour that has been planned out on a map can be perfectly driven, or end in disaster depending on how much time you have taken for planning. In addition to the length of the tour, factors which play an especially important role here are the composition of the terrain and the altitude to climb, which can be found on the map, upon closer inspection. You'll now learn all you need to know in order to prepare for and wrap-up a tour.

Garmin file formats: GPX, GDB, FIT, TCX, CRS

GPX file

A GPX file can contain one, or several waypoints, routes and/or tracks. It is a very universal format that can now be opened by most GPS programs on the PC. Tracks, routes, waypoints and geocaches can be processed in the '64 series device only when in this format.

GDB file format

This is the in-house file format of a Garmin database, which the Garmin map programs on the PC, such as the BaseCamp software, use for building the maps. Files in this format contain everything you have created/edited within a project. In BaseCamp, the links assigned between tracks and photo files can be saved. However, such software-related formats can be reopened only by the same software. Files in GDB format do not belong in the GPS64, as it cannot do anything with them.

FIT, TCX and CRS file format

Next to the normal GPS information, a FIT, TCX or CRS (earlier version) file contains training information from a Garmin training device. Such a file can include one or more stretches, trainings, activities, as well as user/bike profiles and pulse/power/speed ranges.

The models in the GPSMAP 64 series cannot do anything with these files.

Create a backup copy of the GPS device storage

As soon as you connect your GPS64 to the PC for the first time, before the first time you even have the chance to accidentally delete an important system file, the very first thing to do is create a backup file of your GPS device storage!

To do this, connect the device to the PC and wait until the window of the desktop explorer automatically opens. Then, open the device storage with a double click on the drive detected as "Garmin GPSMAP 64..." Copy the "Garmin" folder, which is laying here, to a safe storage device and save this for all eternity. With a "hard reset," (Chapter 2 / "... shortcuts") you can reset the GPS64 to the factory settings and therefore reproduce inadvertently deleted files, but sometimes that does not work. Doing this may potentially save you from having to make a shipment to Garmin.

Device storage: System and folder structure

In order for the files in the device storage to be able to be read, the illustrated folder structure is required.

After you connect it to the PC via USB cable, the device storage of GPS64 is recognized as an external drive (mass storage) named "Garmin GPSMAP 64..." Within lies the "Garmin" folder with critical system files of your GPS device, as well as further sub-folders for their own distinct use.

The sub-folders and their meaning:

- In the "BirdsEye" folder are the satellite and map images, which you can buy online and load into the device. When delivered, there are some demo images here, e.g. Paris.

- Use the "CustomMaps" folder for your personally created map images (e.g. a scanned map of the area).

- In the "Custom Symbols" folder you can store your own symbol icons for waypoint markers.

- Above all, your most important folder is the **"GPX" folder**. In here you store your tracks, routes, waypoints and geocaches in GPX format (if you do not want to use the BaseCamp software for transferring GPS elements). In addition herein also lies the "Archive" folder in which Archived tracks are found, and the "Current" folder, in which the currently recorded track is located – in other words, the current track memory. The "NAV" folder is created by itself, when you use the Route Planner in the GPS64.

- In the "JPEG" folder, you can store photos that contain GPS information, in order to use them as navigation objects via the device's photo viewer, or the "FIND" key.

- You can create the "POI" folder yourself, in order to use additional POI sets, such as Christmas markets, bike rental stations, amusement parks, etc. These must be in GPI file format. In the GPS64, you'll find your saved POI collections via the "Find" key > in the "Extras" category. If you only have a POI collection available in GPX format, you can use the POI Loader

4–107

from Garmin to convert this collection into the required GPI format and to activate the proximity alerts at the same time (see track with turn-by-turn instructions).

You can freely dispose of the previously described folders. Even the deletion of these folders does not affect the operation of the GPSMAP 64. Deleted folders are automatically recreated when the unit is turned on again. The device itself will execute the appropriate action. Your personal files will nevertheless be removed.

If necessary, you can recreate missing folders yourself on your PC using the desktop explorer: right mouse click > New > "Folder."

→ However, in the Garmin folder itself are individual system files and folders that you should absolutely leave as is. Do not delete any files, if you have any doubt about their content! ←

MicroSD card set up

As already mentioned, you can expand the storage space of your GPSMAP64s and st model with an empty microSD card. The devices are currently able to handle memory cards up to 32 GB and a maximum of 2025 map tiles. Use only microSD, or microSDHC cards. Cards with the "ultra II" extension can possibly cause problems, but also offer no advantage in the GPS device, since it depends on the ability to read the map data stored on the microSD card.

In order for the memory card data to be recognized by the device, a folder named "Garmin" must be created on the card (right click: "New" > "Folder"). To put tours and waypoints on the microSD card, it is necessary to have a subfolder named "GPX," in the newly created "Garmin" folder. You can configure this yourself, or have it automatically done by the "BaseCamp" map software, the first time something is sent to the microSD card.

Procedure in BaseCamp: Open any track/route/waypoint stored on your PC's hard drive (File > Import into 'My Collection') and send it to the microSD card (right click on track name in the left column >

"Send to..." > select the SD card). Thus, BaseCamp creates the appropriate folder structure on the microSD card. The same happens when you send map tiles to the microSD card for the first time.

Practical experience: along the way, we have learned that the best functionality and operational readiness occurs when you only send maps parts to the microSD card and all tracks, routes, waypoints, etc. are placed in the "GPX" folder of the device storage. If you are on a trip and need additional maps in the device, you can send this map data to the device storage. This way you will not overwrite the map sections that you will still need again later, when you return home from the trip and can easily delete the travel maps files (**travelmap*.img*) from the device storage. Simply click on this file in Windows Explorer and press the "Delete" button.

Map installation

Depending on the map type you are using, there are different things to do. Let's start with the least amount of effort:

Preprogrammed map data - microSD card

for use in the GPS64. It's simple: Open the battery compartment, remove the batteries, push the metal retaining leaf underneath with a fingernail, open, insert a microSD card with the contact side towards the display, close and slide it back into the locked position. Naturally, reinsert the batteries and close the cover.

The map information on these pre-programmed SD cards can now be used immediately. There is no activation needed. If the display in the map still remains empty, it may be possible that the map is turned off in the device settings ("Disabled"). Please check the Main Menu > Setup > Map > Map Information > select the desired map, confirm and select "Enable" from the menu which appears. Please always leave the Basemap enabled!

If you move around in a border area of this particular map's coverage, it will become unbearable to constantly have to exchange the adjacent map's card manually. In this case, the variant of transferring the map from the PC to an empty microSD card is better, because you can load excerpts from several different Garmin maps onto this memory card and the display perceives no difference at border crossings. The next adjacent map is immediately visible.

For this purpose, the map data, including the corresponding software, must be installed on the PC, which will also require the inevitable activation process via an Internet connection.

Map data - DVD installation on a PC

For this, the following sequence is recommended:

1. Set up a Garmin account. On the homepage of Garmin: www.garmin.com, find the "Sign In" link. There, you can create an account, via the "Create One" link. Register your GPSMAP 64 on the "Your Account" > "Your Products" page and then Sign Out of this account again.

2. Insert the DVD into your PC and start the installation. The wizard unambiguously guides you through the process and at the correct point, prompts you to connect the GPS device to the computer via USB cable, in order to unlock the map for your PC and your GPS device online.

In this manner, all the important registration information can now be credited to your Garmin account. This is advantageous, in that you can access your activation code in case there was an issue during the unlocking process, so that you can more easily order map updates, or check if free map updates are available. This is often the case, when you have bought Garmin Street Maps.

After the installation, it's best to close everything and restart the computer. The recently installed software can then be opened via Start > Programs > Garmin > "BaseCamp." If, when you first open the Garmin software, there is still a pop up message, such as "Map data needs to be unlocked," simply confirm this message again by pressing "OK" and allow the online connection. It shouldn't take more than two openings for everything to be unlocked and to no longer receive the warning. See also www.garmin.com > Support > Software > Show Supporting Software > Mapping Tools to see if a newer software update exists, so that you can best use all functions. In normal circumstances, BaseCamp will update itself, when there is an active online connection.

If only the basic MapSource software was supplied with the road map DVD, you can freely download the BaseCamp software from the aforementioned Garmin website.

You now have software available to you that allows you to perform all sorts of planning and editing features in the map on the one hand and on the other hand, the map itself, which is now immediately available in the software. Should you add more Garmin maps, for example in addition to the "City Navigator North America NT" for the USA, you add "City Navigator Europe NT" for your next vacation, and maybe a topographic map of this area after the successful activation, these maps are all available in the same software. This means you can shovel various maps together into the GPS unit on an empty microSD card.

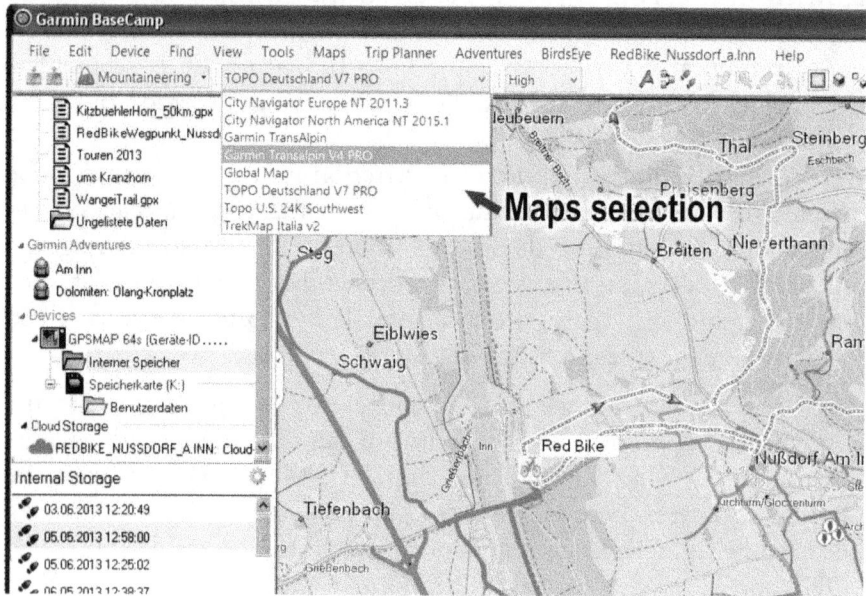

Figure 4-1
The various Garmin maps are selectable in the Garmin Software

Sending maps from a PC to the GPS device

If not already done, now insert a blank microSD card in your GPS64 and connect it to the computer via a USB cable.

Open "BaseCamp" with a left mouse click on Start > All Programs > Garmin.

To transfer maps to the memory card placed in the '64, with the right mouse button click in the left column of BaseCamp on the "User Data" storage folder (which was detected on the microSD card) and choose "Install Maps on Memory Card (...:)" from the context menu.
This opens the MapInstall wizard, from which you can select the desired maps in the window.

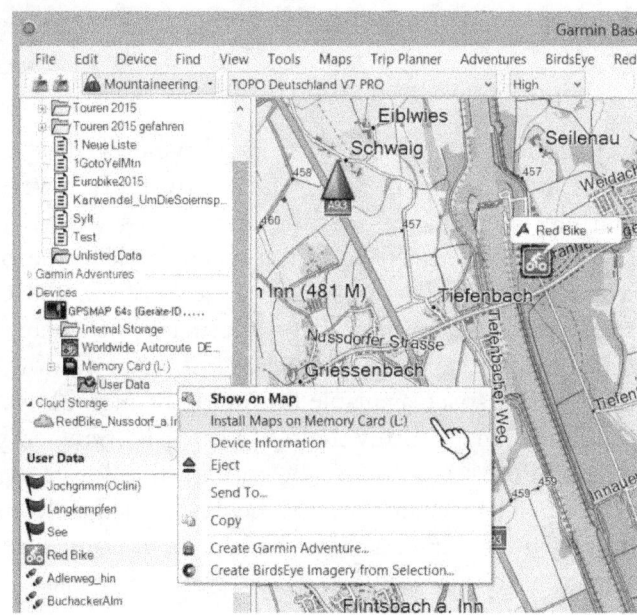

Figure 4-2
Send maps to the microSD card

→ Make sure that you do not add your map collection to a pre-programmed Garmin microSD data card. On the one hand, this no longer has much room for additional map sections, on the other hand the original Garmin map file may be damaged and thus you'll lose the license-free use for other devices. For your own map collection, it is necessary to always use an empty microSD card. ←

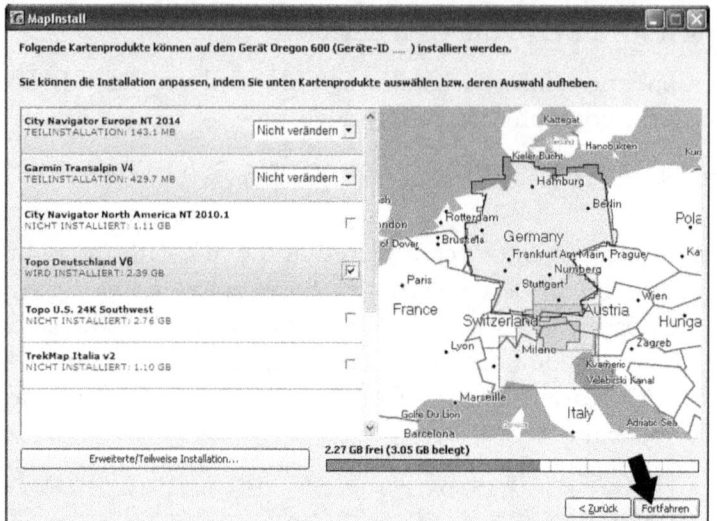

Image 4-3
MapInstall, the "map sending" wizard

On the left side of the MapInstall window, you can see all available maps and their data size. (If you currently have a pre-programmed Garmin microSD card in the device, this would also be visible here, but cannot be selected.) Put a check mark in the box for the map(s), which is/are to be transferred as a whole to the GPS device. In my selection of the entire "Topo Germany V6" map and some parts of other maps, the green bar below the map window signals to me that there is still space available on my 4GB microSD card. So, I could still send more maps to my microSD card. To transfer, you now only need to click the "Continue" button in the lower right corner of the window. The transfer process begins immediately and could take some time (30-60 minutes), if you have a comprehensive digital map selection. In the meantime, go do something else.

However, if you want to only send selected map sections to the GPS device, don't mark any check boxes at all for any of the map rows, but instead, immediately click on "Advanced/Partial Installation" button placed at the bottom left (Figure 4-3).

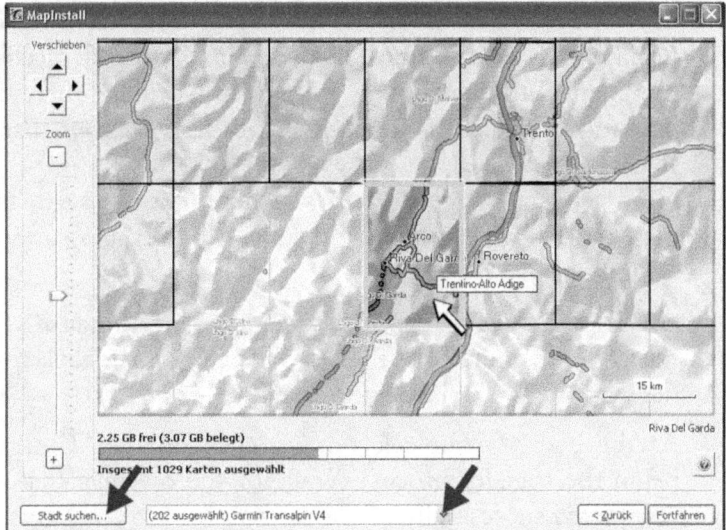

Figure 4-4
Send selected map sections

In the map window that opens, you can now individually select the desired map tiles with the left mouse button, or draw a rectangle over several map tiles, using the left mouse button to select them all at once. To the left of the map, you will find the navigation elements, allowing you to resize, or move the map.

With the "Find a City..." button, found at the bottom edge of the window, you can also go directly to the city, which you want to transfer to the GPS64 along with its surrounding map tiles.

If you have selected the desired sections of one map, now select a second map from which you also want to add some map sections. To do this, with the left mouse button, click on the dropdown arrow of the list that is located next to the "Find a City" button at the lower edge of the window.

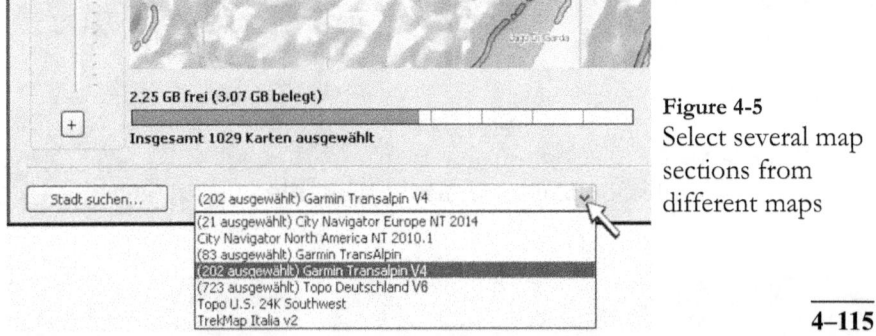

Figure 4-5
Select several map sections from different maps

This opens the list of all the maps that you have installed on your PC. Look closely at the entries in the list and you will discover that next to the map with which you were just working, there is a remark in parenthesis, which is the number of maps sections that have just been selected.

Now activate a second map and here, choose again the desired map tiles (either by clicking individually, or by dragging a rectangle over the desired area).

➔ If you want to <u>deselect a map section</u>, because you clicked it by mistake, press and hold the "Ctrl" key and then click again on this map tile with the left mouse button. ⬅

At the end of your selection(s), click on the "Continue" button to begin the transfer process. Wait for the message that the process has completed successfully.

After the sending process, using the computer Explorer, you should be able to find the map files in the "Garmin" folder of the GPS64's microSD card. The "IMG" files were named automatically, based on the name of each map.

Disconnect your GPS64 from the PC and turn it on. Check in the Setup > Map > Map Information to see if all of the previously sent maps are listed individually. It should be similar to the image on the right. By focusing on the map and pressing "ENTER," you can <u>enable</u>, or <u>disable</u> the respective map in the device, depending on which map is to be used for navigation.

Enable - Worldwide DEM Basemap,NR

Enable - City Navigator Europe NT 2014

Enable - Garmin TransAlpin

Garmin TransAlpin

Disable - Digital Globe

Figure 4-6
Setup > Map > Map Information = Enable, or disable the map

Tours from the "Net"

Pure leisure enjoyment is certainly the option to follow a track, which has been already experienced and recorded live by someone. As in the example of our own GPS download pages:

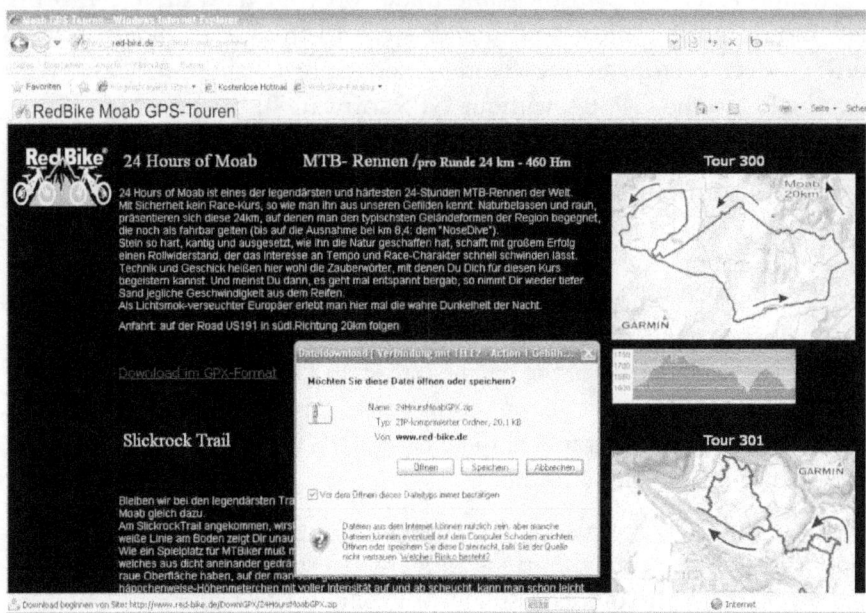

Figure 4-7 GPS Tour portal www.red-bike.de/gps

Here, you will find our own elaborate GPS data that we're crazy about (in the southern foothills of the Alps and some popular tourist regions) with more than 7,000 predominantly mountain bike kilometers (4,300 miles) with more than 200,000 meters (646,000 feet) of altitude. They are all available to freely download for private use.

On these download pages, there are always helpful items, such as a brief description, map, elevation profile and (in the above Figure, hidden behind the download window) the link to the free version of Google Earth. From the many details, this gives you the ability to get an accurate impression, before you decide to download.

For an uncomplicated download, we have packaged our GPS data into a ZIP folder. After saving this to your hard drive, you'll have to unpack/uncompress it with a right-click of the mouse button (select "Extract All"). After unpacking, there is a file, e.g. "ExampleTrack.gpx," in the same folder on your hard drive. With the left mouse button, click and drag this file to the GPX folder on your GPS64-device storage (see figure description: Chapter 3/"Send data from a PC to the GPS64, without GPS software").

Other GPS download portals offer a variety of different file formats, in which you can ultimately download to a track. Garmin outdoor units, and most GPS programs, work with the universal GPX format. Once you have found and clicked the download link for the GPX format, a small dialog box should always open where you can select the location to save. If instead, a new, big, white page opens with infinite number codes, the website designer hasn't correctly configured the download and you would need to know some tricks to still be able to use the track. But with the advanced, digital age of GPS, these types of errors should hardly be still be found.

With some tour portals, such as www.gpsies.com, one also has the opportunity to send the track directly to the device. To do this however, the PC requires a small tool (plug-in) in order to be able to establish communication between the portal and GPS64. This "plug-in" will be installed automatically, as soon as you initiate the GPX file download and grant permission to the tool in the emerging alert.

GPSies.com is an extensive GPS tour portal for all tasks related to the use of your GPS receiver. Here you have the option to do the following:

- To draw tracks online. This is very useful if you don't have any map material available on your own PC for the scheduled tour;
- Convert GPS files with their in-depth transformation program. This is necessary, if you draw a track on any number of

arbitrary electronics maps on the PC, but the software doesn't allow any possibility of storing it in GPX format (for the Garmin device), because every map manufacturer uses its own file format. Many now recognize the universal GPX format, and at the end of the process, offer the option to save/export the track in this format;

- Upload and publish your own tours

and of course, as already mentioned

- Download tours worldwide.

Another great GPS tour portal is www.gps-tour.info, which also offers an extensive, global GPS tour selection. Additionally on the websites of all tourism organizations, there is an increasing number of GPS data appearing for tours of the region, free to download.

Don't blindly trust the provided material. Sometimes they have only been drawn, but in reality never tested. It is also often the case that they haven't subsequently edited (cleansed of errant turns, or direct routing). It is therefore advisable to first always view the files on the PC in your own GPS software, possibly rework them and only then transfer to the GPS device.

In exceptional cases, routes are available for download. But be careful: Because a route is never a recording from a GPS device, it can only just be the planning work of the person offering the route. Once on the road, one runs the risk of not being able to exactly follow the recommended tour, because the route is of course recalculated by the map you're using on your GPS device (otherwise you would just see a straight line to the next waypoint). This is something that is currently rarely found, however, on such platforms. Most places offer their real, previously traveled and recorded tours in the net. This is basically a track with a lot of track points (hundreds, often even thousands of points).

At GPSies.com it is now also possible during the download process to convert a track into a route and download.

Avoid this! Because the track points are converted into intermediate targets for a route and displayed in the form of waypoints on the map. First off, you would barely be able to see the line of travel in the map, due to the sheer number of waypoint flags and secondly, you unnecessarily and uselessly encumber the device with so much computation, that is would be no surprise when it refuses to work and turns off. This is because a route is automatically calculated by the device software, based on roads and paths and never needs as many via-points as a track has track points. Incidentally, the GPS64 only allows 200 waypoints per route anyway and will therefore not even start the calculation.

→ Effectively:
Downloads which have been provided as a track, should not be converted into routes! Because there is no longer assurance that you will be following the original itinerary. ←

Planning and drawing tours yourself

If the very first thing you do is to consider where you want to go on your next tour, then "Google Earth" should probably be at the top of your list. This is available as a free version (http://earth.google.com), which will completely satisfy all private matters surrounding the planning and collecting of GPS tours. Especially for multi-day outings, you can give yourself a very good, initial overview, e.g. a region suitable for an active holiday, but also where some time on the beach should not be left out. You can zoom in to every hotel and have a look at each to see whether mountainous hinterland is nearby. At the same time, you encounter some photos, which display the environment even better.

For drawing a track, Google Earth is only conditionally suitable, since the paths can barely be seen consistently in the satellite image - in reality, it can quickly come to the fact that the supposed path is a riverbed, or it's swallowed by the forest anyway. In high altitudes you can get rather lucky and easily trace alpine footpaths. To create the final tour however, you should use more detailed maps.

Figure 4-8 Drawing a track in Google Earth

If you want to try it in Google Earth anyway (if only to mark a particular area), you start with the "Add Path" tool from the tool bar above the map. This will open the track properties. Here you can now give the track a name, modify the line thickness and change the line color. While drawing, the window must remain open. Therefore, click it at the top and simply drag it down, or to the side, where it does not interfere. Now click along the recognizable path in the satellite image with the mouse, which has been effectively transformed into a pen.

When your track is finished, close the Track Properties window with "OK." The drawn track can now be found in the list to the left of the map. Finally, in order to use this track in GPS64, or further edit it with Garmin software, first save it on your computer: click track in the list with the right mouse button. In the mouse context menu, select "Save Place As..." The only file formats allowed are the Google Earth "KMZ" and "KML" proprietary formats. That doesn't matter. Save your drawn track on your hard disk in "KMZ" format. (KMZ is the compressed form of a KML file.) With the Garmin BaseCamp software, you can easily open the track format and further process, as well as transfer it to the GPS device.

Drawing in BaseCamp

If you are in possession of a Garmin map, you are in possession of a much more secure item with which to do basic planning, as it reveals a lot more path information (without an online connection). You should take another look at all the events from your Google Earth draft with your Garmin map.

When you first open the software, you will probably see the shared map view. One window in the map will be in the 2D view and the other window can be seen in the 3D view. For drawing, you only need the 2D representation. In other words, the view as it is known from a paper map. Therefore, switch to the "2D Map" in the menu bar > View > Map Views. If you are not able to cope without the small Overview Map, you can hide it by clicking through the above described menu tree with the left mouse button and then on the activated "Overview Map." Thus, you now have the most space on the PC screen to quickly, comfortably and clearly create and edit tours in the 2D view.

Click on "My Collection" on the left side in the libraries list to work in your working folder on your PC's hard drive. If you have connected your GPS64 to the PC via a USB cable, the list will include the "Internal Storage" of the detected GPS device and "User Data," which is the inserted microSD card. You can also click either one of these to work directly in this storage medium. So, make sure that you always select the appropriate folder before starting a drawing.

→ By clicking on the "My Collection" PC-working folder with the right mouse button, you can create more "lists" and "list folders" (subfolders), e.g. each list per tour which then includes possible optional routes and important waypoints. In contrast, in the working folder of the GPS device, or its microSD card, no subfolders, or lists can be created. ←

From the dropdown menu above the map window, first select the map where you want to draw your tour (in case you already have more than one map available).

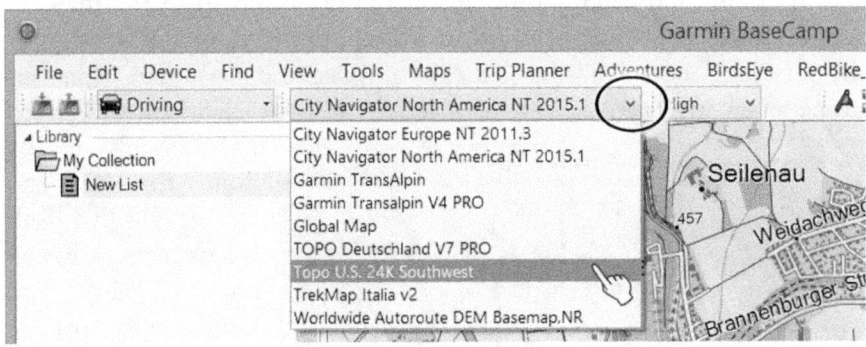

Figure 4-9 Select a map

At the same time, we'll next select the correct <u>Activity Profile</u>. Open the drop down list, that has a picture of a car displayed, and choose your activity, such as "Mountain biking," for example. Now we can use the route function to quickly create a tour. After this, we go into the details.

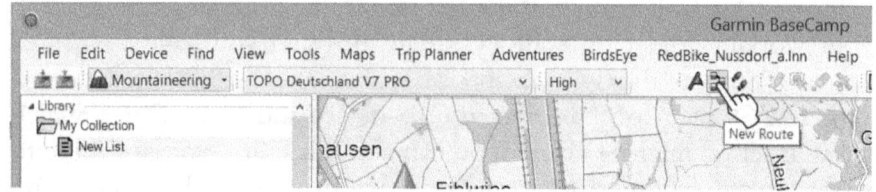

Figure 4-10 Drawing tools for creating a new Route

The drawing tools are in the tool bar above the map. Here, you can select the tool for the creation of a new route. This will open the "New Route" window, with which you could very quickly create a route between 2 waypoints. For this, it would have been required that you create start and end points as waypoints. But since we anyway want to draw our tour in the map, we can just close the small dialog box again and move the mouse pointer in the map. Now place the first click with the mouse, which has now turned into a pen, on your starting point and each additional click along the path in the map, so that the software is forced to create a route per your specifications. Set the distance between your clicks only so far that you can see immediately whether the path between the last two clicks runs as you had imagined. By doing this, you end up saving yourself from having to go back again and redraw every turn.

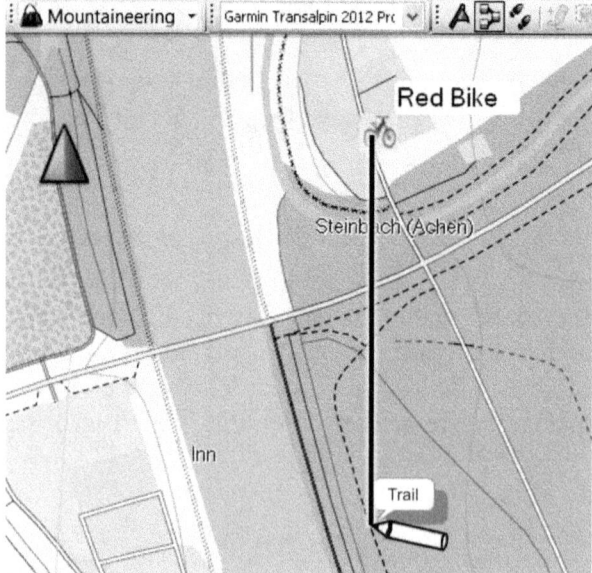

Figure 4-11
Drawing a route

The BaseCamp software calculates the exact route between the clicks according to your selected "Mountain Biking" activity. Draw your entire tour, in this manner. If, at some point, the route was not formed as you had intended, use the shortcut ↶ undo function and set your path marker again. Finish your drawing with a right-click of the mouse.

Subsequently, you can use the "<u>Insert</u>," "<u>Move</u> Point," or "<u>Erase</u>" tools to customize your route design in more detail.

The scissors symbolize the <u>Divide</u> function. If you have designed several routes and would like to combine part of one with a part of another design, you first use this "Divide" tool for separating unwanted sections and then the right mouse button to reconnect the pieces selected from the object list on the left (context menu: "<u>Join</u>..."). Similarly, you can also edit tracks.

At the bottom of the column to the left of the map window, your drawn route is now listed. As we have learned, the resulting route is not based solely on your mouse clicks, but also automatically calculated

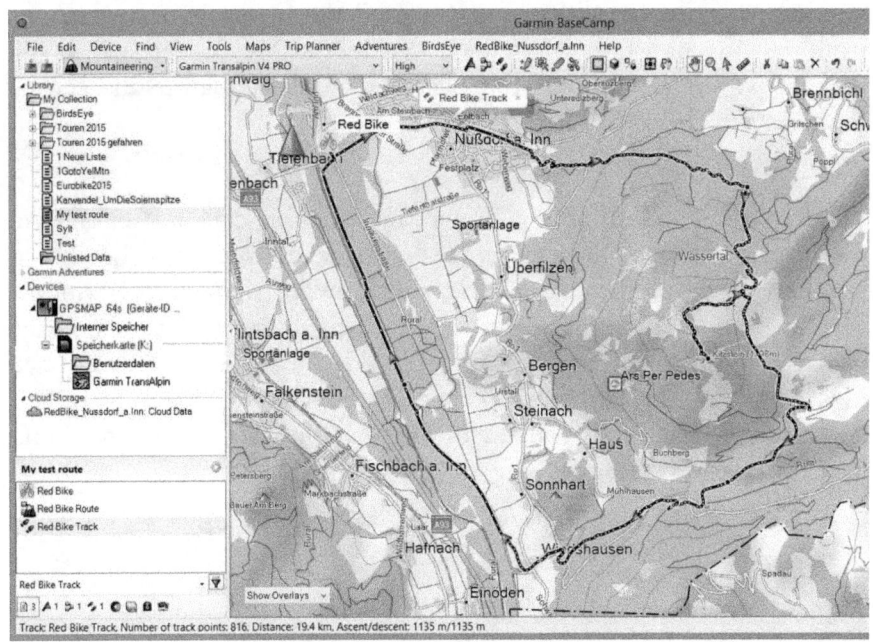

Figure 4-12 Create a track from a chosen route

based on various settings and the basics of map. To ensure that your tour stays the same, no matter what, you should convert your route to a track. Because a track cannot just be automatically changed by anything and you can be 100% sure that the drawn tour continues to be the same in each GPS device and any electronic map on a PC.

The conversion of a route to a track is done easily: with the right mouse button, click the route entry in the left object list and from the context menu, select: "<u>Create Track from Selected Route</u>." The result is another entry added to your object list, which has now got 2 little feet in front of the name, i.e. the symbol of a track. The route thus served us only as a draft, which we now want to delete. In the Windows version of BaseCamp, you can do a lot with the right mouse button.

➜ To completely remove items from BaseCamp, you click on the object you no longer desire, with the right mouse button and select "<u>Delete</u>" from the context menu. There is an almost identical entry. Similarly, if you select "<u>Remove</u> from...," the object is only removed from the current list, meaning the list folder of "My Collection." It still remains within the entire "My Collection" folder structure and can be found in the "Unlisted Data" folder. ←

To learn more about the now-drawn track, double-click the entry in the object list with the left mouse button. This track will be shown centered in the map window and simultaneously opens the window with the track properties. Herein lies the summary data about the distance and forthcoming elevation, provided that the track was drawn in a topographic Garmin map, because in Garmin street maps, no elevation information is contained. The "Graph" tab can also be found in the properties window.

With the help of the elevation profile graphics on this tab, every-thing can be examined in more detail. If you move the cursor along the contour line, this position is simultaneously display-ed in the map window. So, you can get good information in advance about critical ascent or descent situations, especially when it comes to drivability with the MTB.

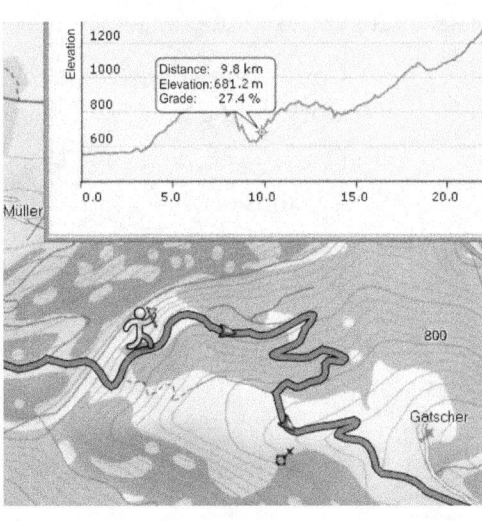

Figure 4-13 The elevation profile provides information on critical cruxes

At the beginning of planning a tour, it is usually the case that the PC screen is too small to see where you have to draw at all. For this situation, there is a simple trick:

First, use the Track tool and simply draw a straight line, on which you can then orient yourself when you zoom into the map.

In step 2, select the Route function again and in the enlarged map view, draw the tour next to your straight line.

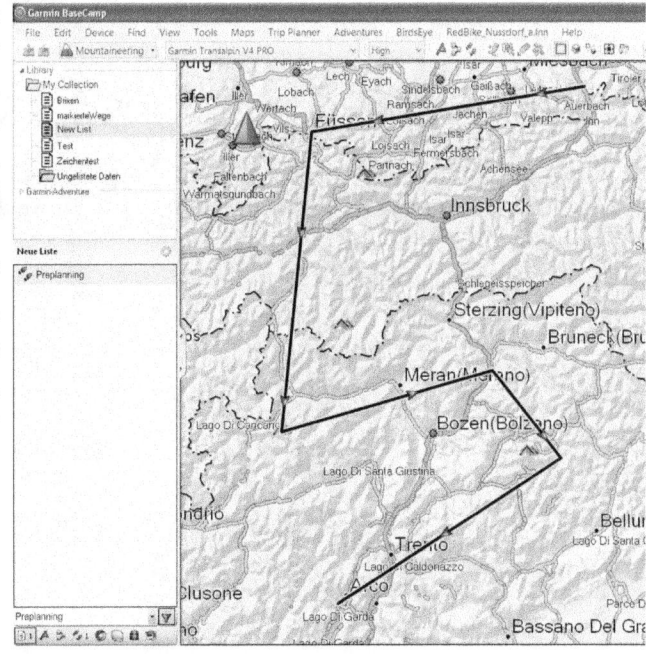

Figure 4-14 Roughly pre-draw the tour

4–127

Don't get mixed up:

- Converting routes to tracks is not an issue and lightens the work during tour planning.

- Converting tracks to routes doesn't make any sense at all – Hands off! –

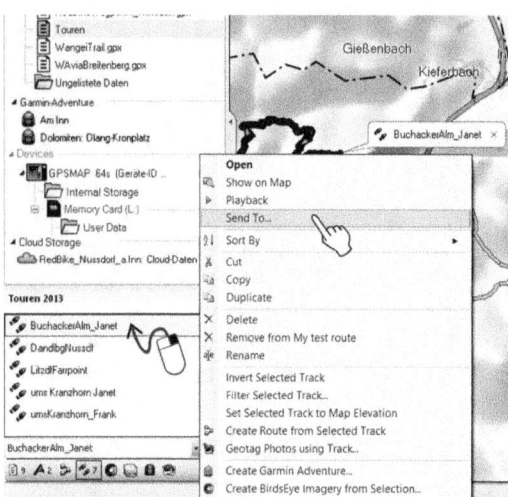

Now we will send the track directly from BaseCamp to the GPS64. Just click the track entries in the object list with the right mouse button again and choose "Send to …" from the mouse context menu.

Figure 4-15
Transfer the track to the GPS device

A small dialog window opens, where you can now decide on the location where the track is to be stored. So either in the device storage "Internal Storage," or on the memory card "User Data."

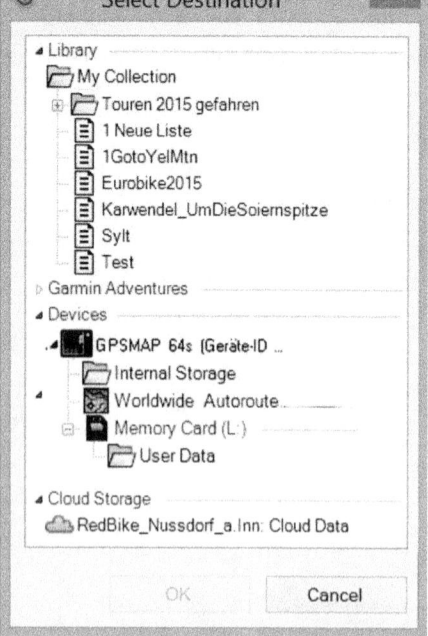

Figure 4-16
"Transfer to" – dialog window

4–128

You can simultaneously send multiple objects to your '64. Just click and hold the "Ctrl" key and with the left mouse button, select all objects that you want to transfer, e.g. waypoints, main track and optional tracks, i.e. everything you need for this tour. Then with the right mouse button, click on the one of the selected items and from the context menu, choose "Send to..." > "Internal Storage" again.

After the sending process, be sure to check whether all objects have arrived to the GPS device. While the GPS64 is still connected to the USB cable, do a pre-control once in Windows Explorer and look to see whether any file has appeared in the GPS64's "GPX" folder. A file named "Track.gpx," or "Waypoints.gpx" should now be there. To keep a better overview, you can also rename everything in front of the ".gpx," via Windows Explorer.

You can also use this "Send To..." feature in the BaseCamp software to copy, or move objects from one list to another. The usual "drag and drop" function also does the job.

The search function in BaseCamp

...has a little bit to offer. At the very least, it provides the typical search-box field in the upper right corner of the BaseCamp software, in order to type a search phrase and then starts the search with the 🔍 magnifying glass button. So far everything's done right!

This opens a "Top results" column, which displays the symbols shown here at the top. By clicking on these icons, you determine which categories should be searched. A click on the icon will enable, or disable this search category. The selected symbol is displayed in a frame. If you move the mouse pointer over an icon, it shows you which type of search would be made.

If you, for example, have the southern German area map section in the map window visible and enter "Sellin" in the search field, every other thing is found, except a city with that name. Clicking on the "More" button, to the right of "Points of Interest," will give you further results, in which Sellin should also be found.

Even more crazy is the search for an address. If I would type "Nussdorf, Am Inn 4" in the search field, BaseCamp doesn't currently come up with the idea to look for a street called "Am Inn" in the town of "Nussdorf."

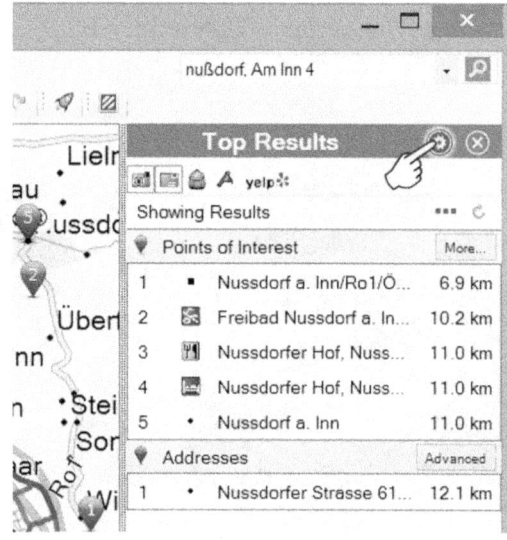

Figure 4-17 Search results: Open Advanced Search Options

The current trick is, namely, to click on the "Open Find Options" ⚙ gear icon in the top line of the "Top Results" column, which opened to the right of the map window, after you clicked search. This will open the Advanced Search Options.

First, you click the checkbox next to "Addresses" and then in the text entry box named "Location," just type "Nussdorf" and you'll immediately notice that a selection list opens beneath that lists all Nussdorfs in one fell swoop. From here, you then only have to click on the corresponding entry. In the initial search field at the top, you now type in the street name with house number and start the search by clicking on the "Search" button.

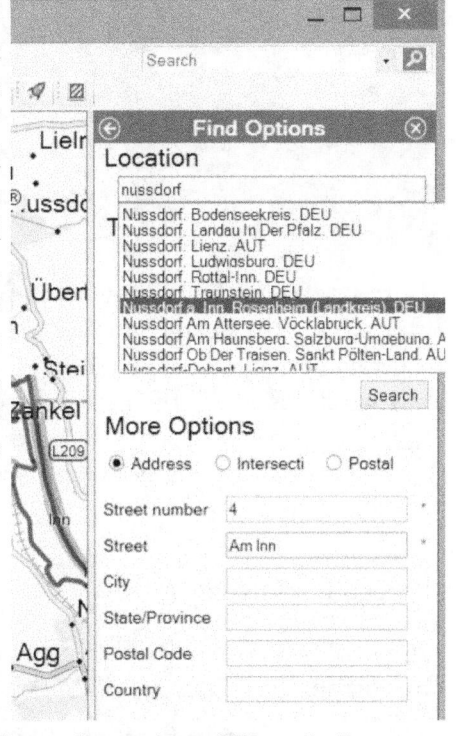

Figure 4-18
Advanced search options

Okay, incorrect, or simply old spelling rules such as "ß" or "ss" go one better. In this case, both must be attempted. You should also keep in mind that the address search with house numbers is only available in road maps, but recently is also possible in the Topo Germany V6 map version.

A much easier way to search is if you know of an area where you think a particular feature exists. If this is the case, choose the ▲ Select tool or the ✋ Move tool from the toolbar on the map window and click your mouse on the map.

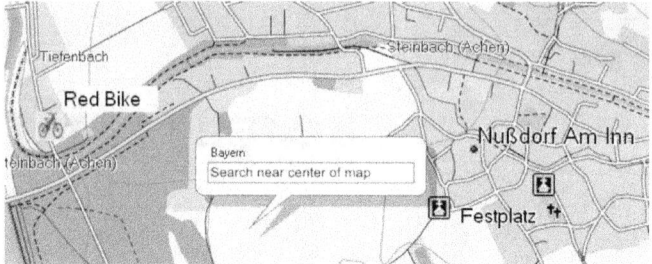

Figure 4-19
Entering the search term in the map window

Pause briefly until an input field opens. Type your search term into it, after which the search immediately starts. Now, you can search for street names, as well as for service providers, since the results are based on such items in the online database of the social networking platform "yelp." Now, you can just type in "Red Bike" here. Thereafter, BaseCamp's own compiled objects will automatically include this with a regional search.

Search for Adventures

In the "Top Results" column, which opens by clicking on the 🔍 magnifying glass button, choose the backpack icon.

That is the Garmin function, which allows you to scour through the Garmin Tour Portal for complete tour proposals, the so-called "Adventures," for the area shown in the map view, or by using the search term entered, while connected online. In the "Top Results" list displayed as a preview, you'll quickly receive finished stretches complete with descriptions, photos and possibly even movies. If you have studied the device chapter regarding Bluetooth applications in more detail, you will recall that we have already accessed this Adventure "collection" via smart device, in order to search for tour suggestions and load them into the GPS64.

Let's now make an attempt from a stationary computer. Enter a term in the search field, e.g. "Kronplatz." As soon as the "Top Results" list opens, select the backpack icon (if not already active). You can deselect

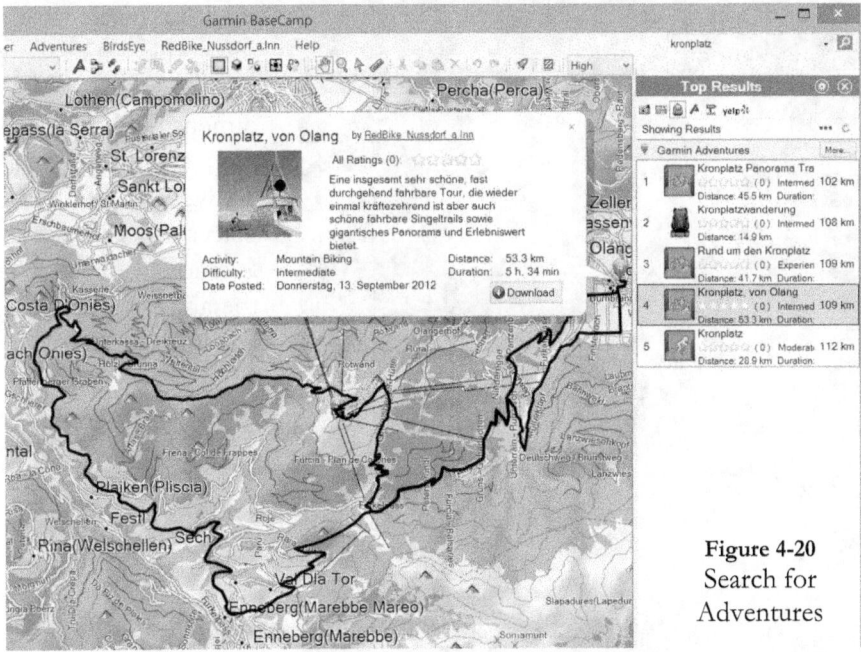

Figure 4-20
Search for Adventures

all other symbols, then the search is faster. BaseCamp already makes an online connection and finds all Adventures that have this text in the name. With a left click of the mouse, the Adventure is displayed on the map, on the other hand by a double click of the left mouse button, the tour will be centered in the map window. A short description opens with details about the activity and the overview summary such as distance and duration. You'll also find a button to "Download" it into the BaseCamp software (Fig. 4-20). If you click on this, you can watch in the left object list as a green progress bar under the "Garmin Adventures" entry appears. You can then open this downloaded Adventure by double-clicking it with the left mouse button. The entire Adventure view opens with detailed information.

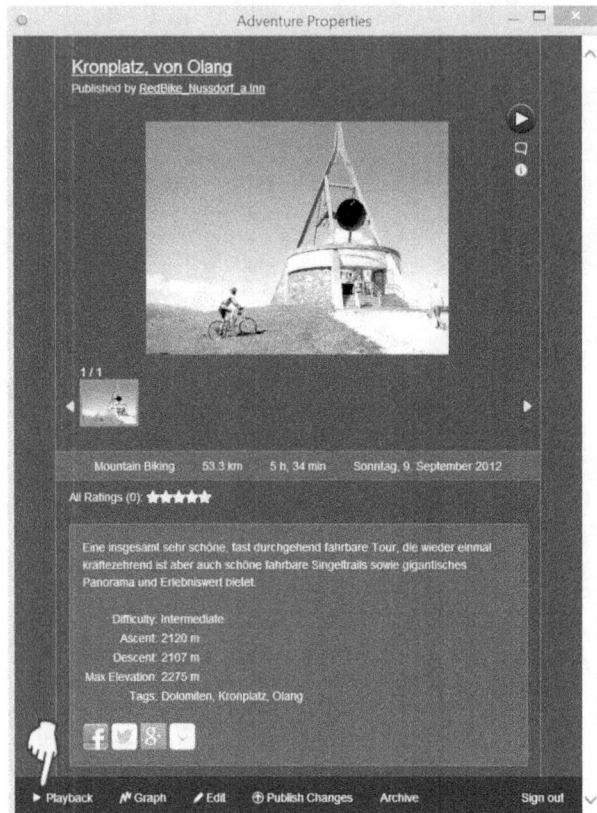

There is also a very funny feature here: the "Playback." With this, the 2D map changes to the 3-dimensional map and starts this "Adventure." In places where there is a photo, or video, the animation pauses for a moment.

Figure 4-21
Detailed information of an Adventure

You can send all data of this tour package to the GPS64 via a transmission process (in the Garmin Adventures object list, click with right mouse button > Send To...)

You start this Adventure in the GPS device by pressing the "FIND" key > Adventures > select the specific Adventure > "Start."

Other planning programs

The information found in a map is not always enough for the planning of a tour. In BaseCamp, you have the advantage that you can switch to the Google Earth application from the current view. With this, you can potentially find some photos showing the itinerary. But in general it is very helpful to be able to have access to more topographical map material.

For example, the KOMPASS map company offers very reliable information in their electronic 3D maps ("KOMPASS Digital Map") with color-coded mountain bike, biking and hiking trails and a clear, familiar map image. These maps have a very good price-to-performance ratio and have a wide range available for common vacation destinations, which should peak your interest. There is, for example, also the typical transalpine area available as electronic version ("Across the Alps" - K 4310). Garmin registers transalpine mountain bike routes with purple and with the detailed way markings of the KOMPASS map, you can very accurately plan bike tours in the Alps and thus largely avoid slide areas.

On the other hand, the software of the "magicMaps 3D-maps" provides very accurate, advanced information about the entire increase/decrease in altitude, the elevation profile and the duration of the tour. Depending on what you enter for your personal travel speeds in flat areas, uphill, downhill, etc. you run little risk of planning too long of a day trip.

Of course, all of these maps also include a 3D function. This gives an even better impression of the terrain features. Often you can even "fly" along the specific track via the animation. It only takes a quick glance at the elevation profile to realize that they are very impressive visually. Drawing is always done in the 2D display, so basically like you would be doing on a paper map.

With time, you will learn the various line types of trails in each respective map, which will help you estimate the suitability of their use, e.g.:

- solid paths = wide forest roads;
- long dashed paths = wide trails;
- short dashed paths = narrow trails: may be drivable with a mountain bike - but likely means a lot of pushing! Then, taking a look at the elevation profile usually gives more information.

With the markings of the path lines in the hiking map, the way the line itself moves (very small zigzag, or winding) and the slope in the altitude profile, it can be seen whether this path is, for example, suitable for mountain biking, or if there are long areas where the bike must be pushed, or even carried.

Anyone planning a multi-day <u>alpine crossing</u> with a mountain bike, but does not want to deal with a lot of route planning, should absolutely take a look at the online tour planner from Uli Stanciu. As the longtime leader of the MTB TransAlp Challenge, he has cataloged popular MTB routes, so that his planning tool even provides accurate information about the road surface. The planned tour can be purchased and downloaded for a few cents per kilometer. All information about it can be found on www.bike-gps.de.

➔ No matter in which software you create the tour, regardless of whether it's a track, a route, or just individual waypoints, at the end, the objects must always be saved in GPX format, or in a format which can be changed to GPX format with the converter on www.gpsies.com.

GPX files are stored in the "GPX" folder, in the Garmin folder of the GPSMAP 64 device, or microSD card storage:

- either through the BaseCamp software (mark the object(s) and right-click "Send To..." > "Internal Storage"), whereby the GPS objects are transmitted, placed in the "GPX" folder and automatically named, or
- with the Drag & Drop function of the desktop explorer, with which you can arbitrarily rename each file and therefore quickly recognize it. ⬅

Elevation values: barometric, via GPS, or from the map

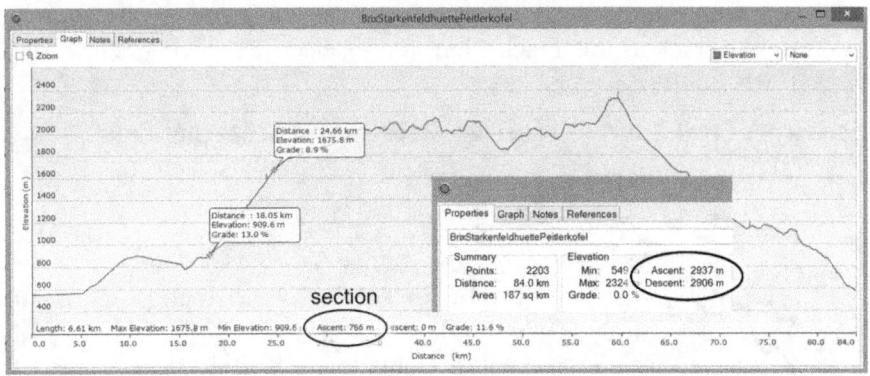

Figure 4-22 Total elevation difference of a Tour

Note: All programs, which give an advanced indication of the <u>elevation difference for a planned tour</u>, only show a theoretical value anyway. During the calculation (summation) of the individual elevation differences that are shown on the drawn track, which might even be less than one full meter, there may be a substantial deviation in the final sum, when compared to the actual forthcoming total elevation difference. Our experience with planning tours has shown that there is usually about 10% more planned, than what you have to overcome in the actual total elevation. You'll have to experience each map yourself, to see how it will truly be. Initially however, just keep in mind that it could also be the other way around.

Therefore, one and the same track can also display different total ascent values, when opened in different map programs.

Even if you open a track in an electronic map on the PC, which was recorded with the GPS64, it could also be that your <u>barometric elevation data</u> from the GPS device is ignored and is replaced by the elevation data in the map. BaseCamp takes the barometric altitude values from the GPS device as its primary source, but you also have the option to manually select "Use map elevation data."

The true result is only shown by the barometrically measured "Total Ascent," which can be read on the GPS device's display in the Elevation Plot and/or Trip Computer view. If you have already saved a track to the unit and have already zeroed out the data field values of your trip computer, you can see the total ascent value of the tour by opening the track in the Track Manager > select the track > View Map > "MENU" > Review Track.

The elevation data created by the GPS is, if at all, only useful when at a standstill. When moving, the GPS-based altitude recording is many times more inaccurate than the barometric recording.

Creating waypoints in BaseCamp

Those who have a spontaneous nature and want to rely purely on the route navigation of the '64, can create only the waypoints in BaseCamp, which can then be quickly selected in the GPS device while underway, because you might not yet know in what order they should be reached. In this way, you can quickly and easily call up the prepared waypoints into the destination menu and start routing to this point with the previously selected correct routing settings.

For the creation in the BaseCamp software, enable the ⚑ flag tool in the toolbar above the map window and with the mouse, click on the point on the map where your new waypoint is to be placed. Then open it by double-clicking with the left mouse button on the new entry in the list of objects. Here, you can assign properties, a name, and

possibly a different icon. In this manner, create all possible waypoints, select them all together in the object list and transfer them via the transmission process to the GPS64 (right click on the selection > Send To... > Internal Storage).
Turn on the GPS64 > "FIND" key > Waypoints > ... > "Go!"

Create waypoints via coordinates

Meanwhile, it's often the case that you will find driving directions specified with coordinates on the Internet, or you have quickly ferreted out a point in Google Earth with the search function. But how do you get this waypoint into the GPS64, if no download link is available?

Example: In the "Search" field of Google Earth, enter a place, or a feature you want to search. Let's say, "Red Bike Nussdorf" and click on the magnifying glass to the right to start the search. In the list below, a selection of points, which are most similar to the search parameters, should then appear. "Red Bike, Am Inn 4, 83131 Nussdorf" should also appear in the list below. Clicking this twice with the left mouse button causes the waypoint to be displayed in the center of the map section. Now activate the "Add Placemark" tool from the toolbar above the map window and the drag the flashing pin directly onto the point which is already shown to you on the map. If necessary, zoom in on the map with the scroll wheel of the mouse as far as required, in order to see the target accurately. In the "New Placemark" window, which automatically opens, you can then read the waypoint coordinates in the latitude and longitude line:

Latitude: 47° 44.511'N

Longitude: 12° 8.313'E

If the format of your coordinate display looks different, you can change it in the settings of Google Earth (in the menu bar: Tools > Options, "3D View" tab > in the "Show Lat/Long" field, select another format. The example shown here is called "Degrees, Decimal Minutes").

Now, open your map software on the PC and create a waypoint. In BaseCamp select the ⚑ flag symbol again in the upper tool bar, then click randomly in the map to create a waypoint, somewhere. This creates a new entry in the object list. Due to the flag in front of the entry, it can be seen that this is a waypoint. Double-click this entry with the left mouse button, so the waypoint properties open. You can enter the coordinates from Google Earth in the "Position" field.

Of course, it's now important that you are using the same coordinate format in your map software as in Google Earth. In addition, the location indication in the Garmin software is written in a single line and without the degree and minutes character. Furthermore, instead of the degree symbol, a space is inserted. The direction information "N" and "E" are written in front of the respective numerical value. The position of the randomly clicked point should therefore now look like the following in BaseCamp:

N47 44.511 E12 08.313

If this is not the case, you can change the position format in BaseCamp as follows: menu bar > Edit > Options > "Measurement" and here in the "Position" field. In the "Grid" drop-down list, select the "Lat/Lon hddd°mm.mmm'" entry. The Map Datum, "WGS 84," remains as is.

In the same way, the waypoint properties can now be supplemented with all information that it should receive, such as the information about the elevation we just discussed. If the randomly clicked point was made within a topographic map, an entry for the elevation is already created.

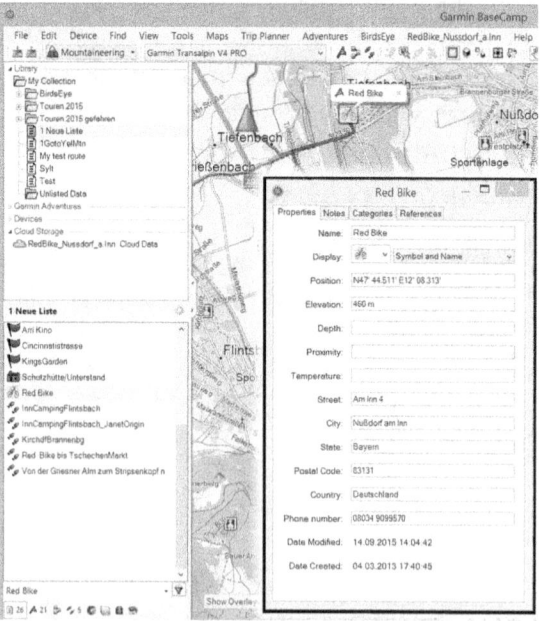

Figure 4-24
Create a waypoint in BaseCamp

4–140

If it was done in a Garmin street map, no elevation entry will exist, because these maps do not contain elevation data. You can now determine the correct elevation by enabling the ![Select] Select arrow in the Map Tools toolbar and move it around the map near the waypoint to look for elevation information. If you point to a contour line, for example, the elevation is displayed. If nothing is found, you could also use the "New Track" tool to start a new track on the point in question in order to display the starting elevation in the elevation profile. Then enter the determined elevation value in the "Elevation" field of the waypoint properties.

All values entered, including the comments that you have written in the "Notes" tab, can then be read in the GPS64, when you select the waypoint in the Waypoint Manager.

Create your own waypoint icon

Those of you who would like to have your own icon as a waypoint on the PC and in the GPS64 can create any arbitrary, well distinguished image with a size of 16x16 pixels and save it in "BMP" (Windows Bitmap) format.

The location for these custom icons on the PC is the "Custom Waypoint Symbols" folder, which should have been automatically created in the "My Garmin" folder under "My Documents" when the software was installed on your computer. If this is not the case, just create a new folder there by right-clicking "New" > "Folder" and name it exactly as shown above.

The name of your image file must consist of 3 digits. It's best to first look in the existing folder to see which image files already exist. Review the number sequencing and then give your image the next higher number in the series, such as 020.bmp, 021.bmp, 023.bmp, etc.

In BaseCamp you'll then find your own icons in the Waypoint Properties, in the symbol selection under "Custom."

To use this icon in the GPS64, copy the image via the desktop explorer to the "CustomSymbols" folder in the "Garmin" folder of the GPS-device storage. In your GPS unit, you'll also find the icon when selecting a waypoint icon under "Custom."

Georeference photos

In the age of GPS and Google Earth, it is a matter of course, that you can also find photos linked to specific waypoints, with which you can get an even better impression of an unfamiliar place.

If you also want to link photos with GPS data from a conventional camera, you can geotag your photos manually by using the free image file editing tool "Geosetter" by Friedemann Schmidt. For downloading and to see a description, visit: www.geosetter.de.

In this program you have the ability to attach countless details to your photo file. If you are not currently an enthusiast of the photography scene, you will probably be downright stunned, when you open your first image and see the wealth of "secret" information that a photo is carrying around with it.

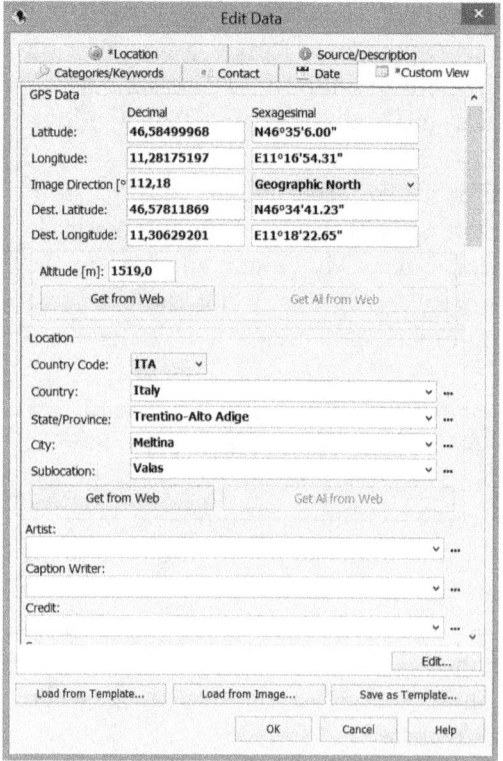

Figure 4-25, above
Mouse pointer on a photo file

After entering the coordinates, everything can be easily supplemented with "Get from Web".

Figure 4-26, left
Properties window of a photo file

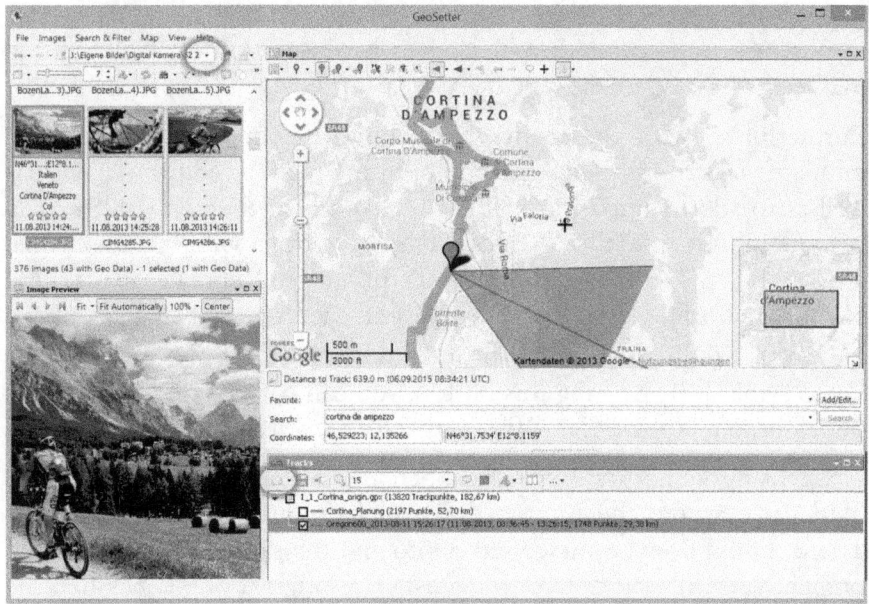

Figure 4-27 Assign coordinates and direction of view to your photos

Once you have the program installed and running on your computer, open your own photo folder by selecting the file path in the dropdown menu, just under the menu toolbar.

To the left of the map, where your photos are now displayed, the best is to immediately double-click on a photo. This opens the properties window with the pre-existing information from the date the photo was taken, which could now be changed. In addition to the Artist, Object Name, Caption, links, contact information, etc., on the "Location" tab, you have the ability to enter coordinates and altitude information, or query these details online. On the other hand, you can also use waypoints, which you, for example, created with your GPS at same location where you took the picture. Doing so makes assigning coordinates to the photo much easier. For this, open your GPX waypoint from the tool bar at the bottom right window. Click the dropdown next to the yellow folder icon. A dialog box appears, through which you can access the storage location and your waypoint and open it in Geosetter. Then, click on the waypoint in the map which

4–143

results in a pop-up window that gives to the option to "Set position marker to focused image."

Furthermore, with the many tools above the map view, you have multiple ways to create a photo point in the map, to edit the details, as well as to add its line of sight. If you have finished editing all the data of your photo, use the "Refresh Files" button with the two green arrows (in the toolbar above your photo collection), or use the keyboard shortcut "F5" to save all the data processed in your photo file. In the toolbar you will also find a button with the blue ◎ Google Earth symbol. By clicking this button, you can convert one, or multiple photos into a Google Earth file, save it, or view it in that program.

Now, using Windows Explorer, open the photo folder on your PC hard drive where the photos that you have just edited are located. There should now be the edited photo and the same photo, but in the original version (without coordinates). The original photo has had the word "original" appended to the JPG file extension, which has rendered it temporarily unusable. Rename your original file and specifically remove the "original" extension addition. Take the edited photo with the GPS information and copy it, using Windows Explorer, to the JPEG images folder of your GPS64-device storage, in order to be able to use it in the GPS device for photo navigation.

Figure 4-28 Use the photo in the image viewer for navigation

Photos, that you have georeferenced and loaded into the JPEG folder of your GPS64 via the desktop explorer, are found again in the Main Menu > Photo Viewer. You can open the photo by navigating to it and pressing "ENTER" and then find further options by pressing the "MENU" key. The selection which appears (Fig. 4-28, 2nd picture from left) gives you the opportunity to see the photo in the map with "View Map" (3rd picture from left), or "View information" to find information about the photo itself (4th picture from left).

In the two latter views, the "Go" button will also appear, allowing you to immediately start the automatic navigation to this location.

Evaluate track recording on the PC

Depending on the software, you can evaluate your GPS recordings on your PC to the smallest of detail, edit and re-use them. While moving, you had the current data of the "journey" in the device display in front of you. Subsequent inspection on a home PC offers a graphical representation of the activity provided as a line on the map, as a terrain representation in the elevation profile, or the 3D view, as well as the tabular listing of all GPS, sensor and movement data.

Open a recording in BaseCamp

So, connect your GPS64 via USB cable to the PC and explore what you have accomplished! In the Garmin BaseCamp software, you can immediately see all the data that is saved in your device and comfortably evaluate and edit it.

To view the recorded data, you only need to click on the "Internal Storage" folder of your '64 (left column) with the left mouse button. You could reprocess the data there as well, or save the raw data into a GPX, or GDB file on your PC. To do this, simply select all, or only the desired recordings ("Ctrl" key pressed) in the object list of the device's storage, and choose File > Export > "Export Selection." So, you don't even have to first send these objects from the device storage to the "My Collection" PC working folder.

If you want to get the data into your PC's working folder, however, you can send a single recording from the GPS unit to the BaseCamp library. You use the "Send to ..." entry from the context menu of the right mouse button, which you already know. First, create a separate list in your "My Collection" library folder. Then click with the right mouse button on one, or more recordings in the object list of the device storage and select "Send to ..." In the "Select Destination" dialog box that opens, choose the list you just created. Doing this allows you to edit the track separately from all other elements, add waypoints from the unit and not lose oversight. You can also use the filter options at the bottom of the Object Browser to display so only tracks, only waypoints, only photos, etc.

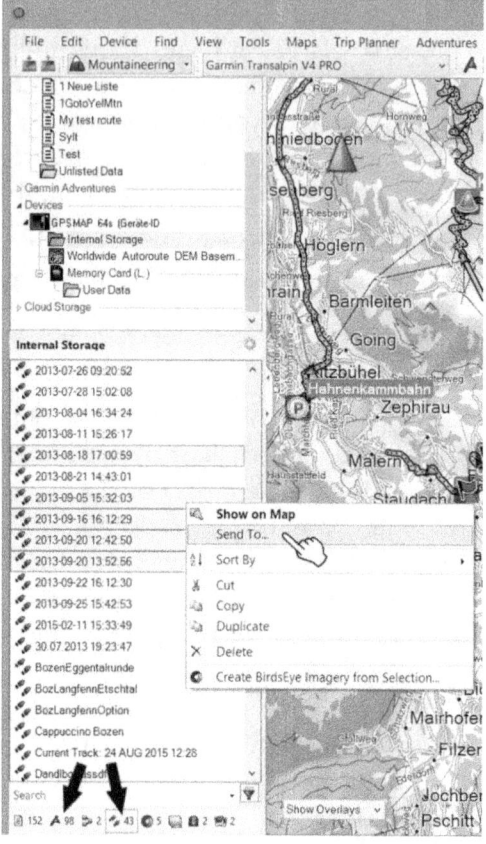

Figure 4-29 Send files from the device memory to a list, or list folder

On the other hand, you can also utilize the 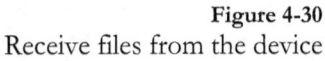 "Receive from Device..." button. Then, the small "Select Device" dialog window opens, where you can select the external device whose data you want to receive. Click on the appropriate line and confirm your choice with "OK". In the library, a folder list is created and automatically named "Data received from GPSMAP 64..." and all the objects from the device are copied to here.

Figure 4-30 Receive files from the device

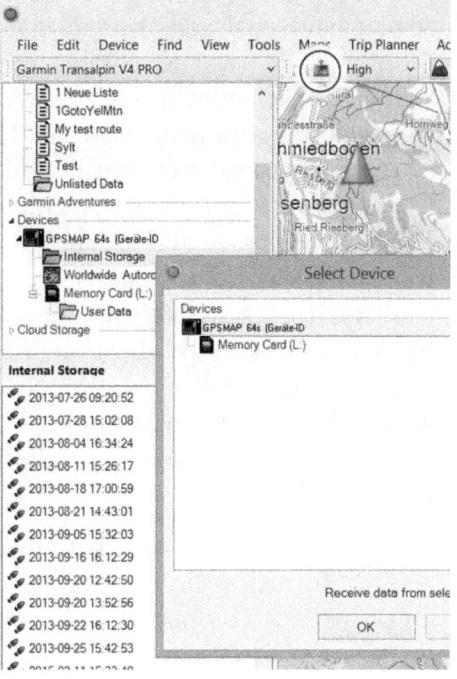

Figure 4-31 Show curves of additional sensors, air temperature curve only possible with the optional "tempe"

In the object list with a double-click of the left mouse button, select a track that was recorded by the GPS64 and transmitted to the "My Collection" working folder on the PC, so that it is centered in the map and will be displayed as large as possible. At the same time, the properties window opens. On the Graph tab in addition to the elevation contour line, you can also display the graphical history of speed and other available sensor values like pulse and temperature. If you move the mouse pointer along the line, you'll get the position displayed on the map.

4–147

If you have analyzed your track experience in detail, close the track properties window and devote yourself to post-editing. In order to now "erase" errant paths, to cut the track into pieces (in case you forgot to stop recording immediately at end of the tour and save it), or to also insert track points to eventually add a small detour, use the editing tools from the toolbar on the map, with which you are already familiar.

Finally, you'll save this cleaned-out, completed post-processed track with the associated waypoints, etc., for eternity. It would make sense to store each track individually including any associated waypoints (e.g., parking spot start of the tour, beautiful views, places to get water, cultivated pastures) in a GPX file on the PC hard drive, or wherever, with a clearly identifiable name, for example, "ParadisePointTrail.gpx" (select the corresponding list in BaseCamp, menu bar: File > Export > Export 'List Name').

In principle, it does not matter whether it is stored in the Garmin "GDB" format, or in the universal "GPX" format. But if there are a lot of extensive properties and possibly also links with photos in a file, in this case you'll have to use the "GDB" format in order to again open this file in BaseCamp, with all its objects and possible links to photos. (The photos, however, must then also lie in this location.) If only the route with waypoints needs to be stored, then the universal GPX format is certainly better, because you can then also use this file in the GPS device, as is.

If you want eventually view a track, or route, as a running animation in the 2D, or 3D map, BaseCamp provides a playback function for this possibility. To do this, activate "Playback" via View > Toolbars in the menu bar and in the playback panel that emerges, select a playback speed between 0.5 to 1000 times the real recording. Now double-click a track, or route from the object list, so the properties window opens and switch to the elevation profile. Drag the window to a point where it will not bother you if you want to see the track, or the route on the map simultaneously. Finally, start playback with the play button in the playback panel. Thus, your position will now be displayed simultaneously on the map and elevation profile.

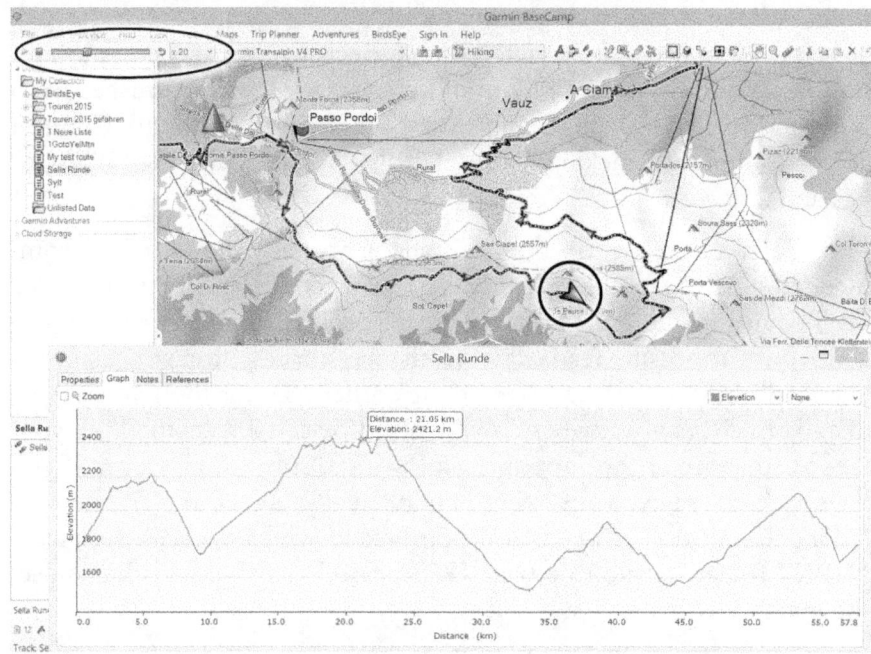

Figure 4-32 Running animation with a speed of 20 times faster than the original recording speed

Open a recording in Garmin Connect

In Garmin's training portal for sports enthusiasts, "Garmin Connect" (http://connect.garmin.com), you have the option of viewing and managing your GPS recordings online and in particular to evaluate the fitness values in somewhat greater detail.

The free Connect web service is also especially useful when you have no maps yourself, because Garmin Connect uses satellite images from Bing and Google Maps for its map backgrounds. With the Connect account, the Live Tracking function is also possible with the GPSMAP 64s and ...st.

Procedure: (The use of the "Google Chrome" browser is recommended.) After you have created a user account and connected the GPS device to the PC, it will be automatically recognized by Connect and you can easily read data from the device storage using the "Upload" button (in the upper right corner). Using the two buttons "Upload All New Activities," or "Upload Selected Activities" all your records that are in the current track log are displayed. However, if you have already saved your recording manually, select the 3rd button "Manual Upload." It opens a dialog box in which you can select the "GPX" folder location of your GPS device's storage and then select the track file.

On the Dashboard tab, you will find an overview of the most recently uploaded tours, as well as your personal best (it is a training portal, after all).

On the Analyze tab > Activities, you can see all your uploaded GPS tours. By clicking on the respective tour, you will be taken to details with elevation information, speed, etc.

➔ But again, beware of the details of the total elevation difference: In the lower left, the details of each record are displayed. They show whether the highly accurate elevation data were taken from the GPS64 with the barometer (elevation correction disabled), or if the offset data of a survey service was used. ←

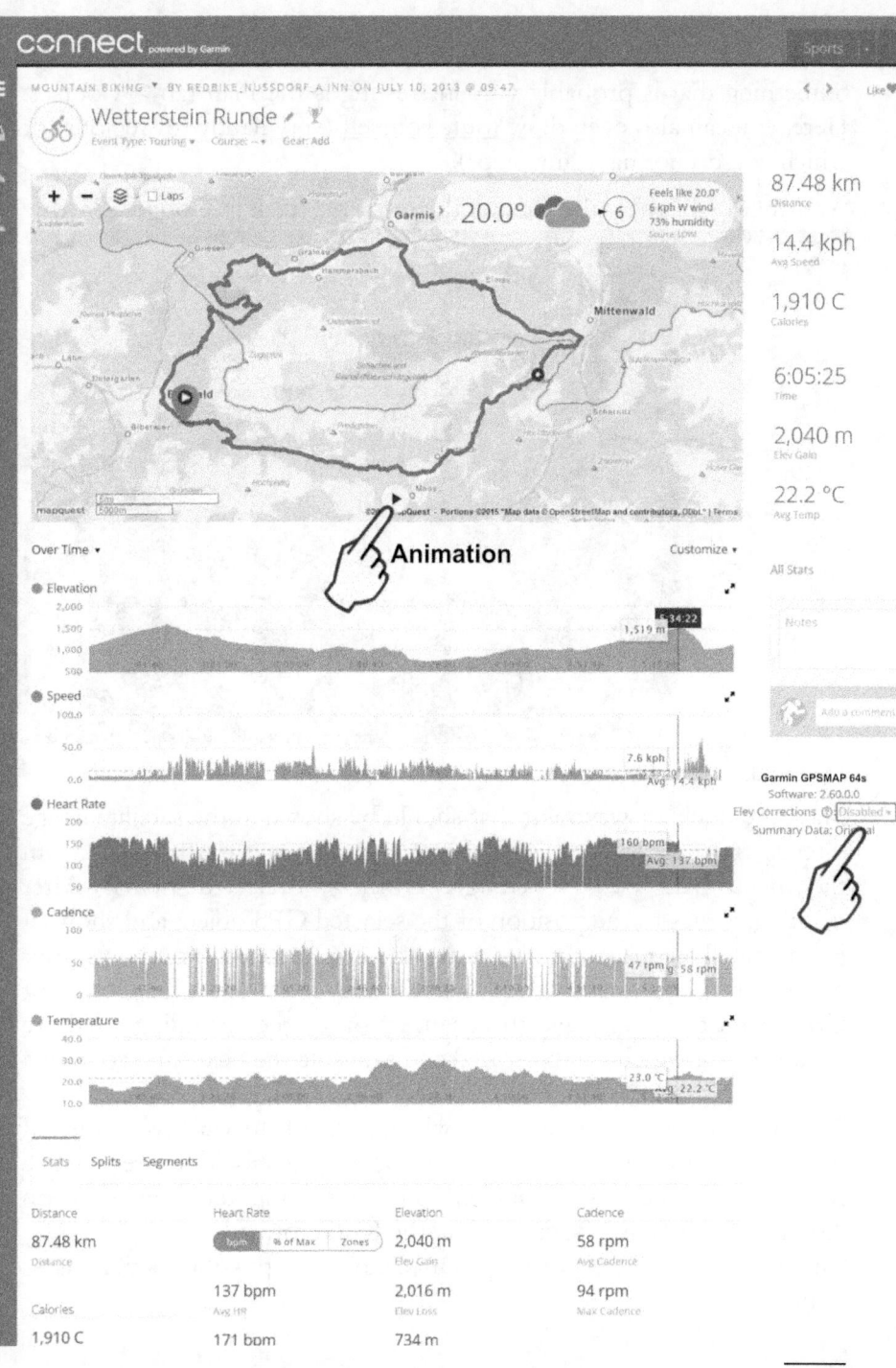

Something that is probably very interesting is the Plan tab > Courses. Here, you can also even draw tours yourself (very handy for regions for which you do not have any maps).

A further highlight is the view of the recorded track in **Google Earth**:

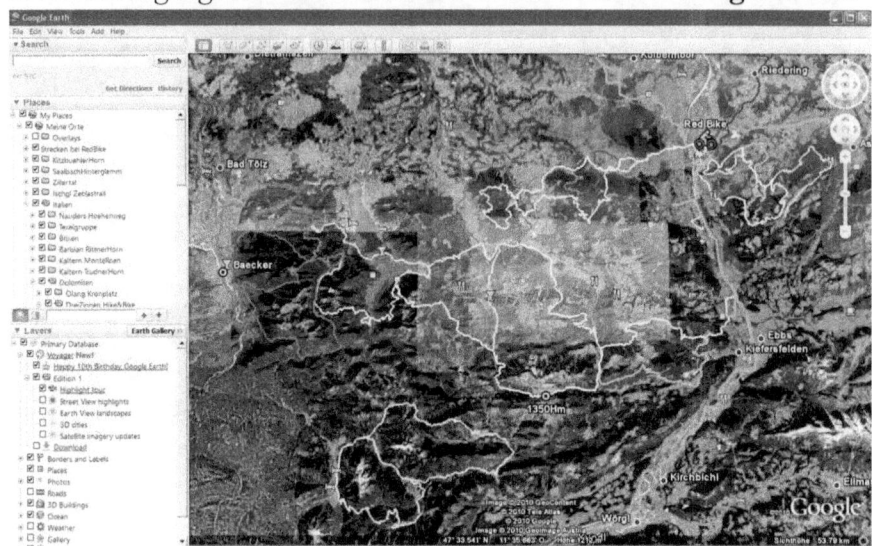

Figure 4-33 An overview of your tour in Google Earth

Mark the track, or waypoint that should be shown in the satellite image directly from the object list in BaseCamp. Google Earth opens from the menu bar: View > Google Earth > "Selected Items" After opening, it flies to the position of the selected GPS object and shows it in the satellite image. In this way, you can generate your own tour portal with very clear overviews. At the very latest upon the closing of Google Earth, you'll have the chance to save the recordings that are already open, which can be stored and available the next time you open Google Earth.

This way, you can see at a glance which parts of the earth you yourself have experienced live. Every now and again you should, however, save the entire collection of tours on your own computer (right click on "My Places" > "Save Place As..." in the KMZ format).

This feature also helps during tour planning to possibly better assess critical points.

So you see, there are a number of options and tools with which you can evaluate and edit the recorded GPS data on the PC, as well as properly deploy for reuse.

However, before passing along your data, think about whether the recorded track is really intended for the public! Avoid the following:

- recording of private roads on which all traffic is undesirable, or even forbidden,
- narrow trails that are possibly even more heavily frequented as MTB routes and
- recording of off-road activities.

Be sure to maintain a healthy balance between man and nature, with all of the different activities it has to offer and have fun with your model of the GPSMAP64 series!

Index

A

Achive folder 4–107
Active Route view 3–94
Active Route, Page Sequence 2–60
ActiveRouting 1–13
Activity Profile, BaseCamp .. 4–123
Activity, routing settings 3–81
Adjust Zoom Ranges 2–51
Advanced Find Options 4–131
Advanced Map Setup 2–55
Adventure 4–132
    Download 4–133
    Start 4–134
Alarm at Waypoint 3–95
Alarms 3–96
Alpine Club maps 1–13
Alpine crossing, MTB 4–136
Altimeter Setup 2–62
ANT+ technology 1–11
Application
    BaseCamp Mobile 2–72
    Garmin Connect 2–69
Application missing 2–61
Archived Tracks 3–101
Archiving, automatic 2–58
Auto Calibration
    Altimeter 2–62
Auto Zoom function 2–55
Avoidance Setup 3–81

B

Backlight Timeout 2–42
Backpack icon 4–132
Backup file 4–106

Barometer Mode 2–62
Barometric elevation data 4–137
BaseCamp
    send maps to the GPS 4–113
BaseCamp Mobile App 2–72
basemap 1–11
Basemap, activate 4–110
Batteries 2–41
Battery level 2–42
Battery Type, settting 2–53
Beep, routing 3–79
Birds Eye Satellite Imagery 1–14
BirdsEye folder 4–107
BirdsEye Select 1–14
Bluetooth 2–68
Bluetooth, transmission 2–72

C

Cadence sensor, couple 2–62
Calculation Method 3–81
Calibrate Compass 2–57
Carpool Lanes 3–81
CDI mode, compass 2–57
Change Data Fields 2–64
Change Item Order 2–61
chirp 1–30
Clear Track and Trip Data 2–59
Cloud 2–74
Colorsettings, track in GPS ... 3–93
Compass scale 2–56
Compass, automatic setting ... 2–57
Compass, calibration 2–57
Connect 4–150
Coordinate entry 3–82
Coordinate system 1–34

Create custom icon 4–141
Create waypoint icon 4–141
CRS file 4–105
Current folder 4–107
CustomMaps folder 4–107
CustomSymbols folder 4–107

D

Dashboard, Change 2–63
Data Field, large 2–63
Data Fields, change 2–64
Data Fields, device 2–54
Data fields, set up 2–63
Data filters 4–146
Degrees Celsius 2–52
Delete
    Current Track 2–59
    Track, all or individual 3–102
Delete a waypoint 2–43
Delete map, device 4–109
Delete Waypoint 2–45
Delete, BaseCamp 4–126
Demo Mode 2–53
Destination entry, Route 3–82
Detail level, map 2–55
Device files, select 4–147
Device storage, track points .. 2–58
Direct-Routing 3–81
Distance to Dest. 2–64
Divide, BaseCamp 4–125
Download, Adventure 4–133
Draw tours, Connect 4–152
DVD, install 4–111

E

Edit Route 3–85
Elevation correction on PC. 4–150
Elevation data

barometric 4–137
per GPS 4–138
plan prior to tour 4–137
Elevation Plot, Route 3–85
Elevation Profil, modify 2–48
Elevation profile, usage 2–50
ENTER key 2–46
Enter text literally 2–44
Erase, BaseCamp 4–125
ETA at Destination 2–64
Express 9

F

Factory reset 2–59
Factory settings 2–55
Find a City, map installation 4–115
FIND kez 2–46
Find Options 4–131
FIT file 4–105
Fitness accessories, activate ... 2–62
Folder structure 4–106

G

Garmin Connect 4–150
Garmin maps 1–12
Garmin Serial 2–53
GDB file 4–105
Geocache
    create filter 1–31
    filtering 1–30
    logging, found 1–29
geocache information 1–27
geocaching 1–26
Georeference photos 4–142
GLONASS 1–38
Google Earth, program 4–120
GPS reception 2–42
GPS Software 1–20

GPS-Download Portal..........4–117
GPX file...................................4–105
GPX folder..............................4–107
GSC 10......................................1–11
GSC10, activate.......................2–62
Guidance Text.........................2–54

H

Handbook.................................2–41
Hard reset................................2–48
Hardware, remove..................3–90
Heading, device setting..........2–56
Heart rate strap, couple..........2–62
Highest, lowest points............2–60

I

IMG map file...........................1–17
Insert, BaseCamp..................4–125
Interface...................................2–53
Interim destination, enter.......3–84

J

Join, BaseCamp.....................4–125
JPEG folder............................4–107

K

KMZ and KML files.............4–122
KOMPASS maps...................4–135

L

Language..................................2–52
Lap time...................................2–63
Large numbers.........................2–63
LAT/LON................................1–36
Line color, track......................2–58
Live-Tracking..........................2–69
Lock Direction........................3–98

Lock On Road.........................3–81
Low points...............................2–60

M

MagicMaps 3D-maps............4–135
Magnetic...................................1–37
Magnetic fields, compass.......2–57
Main Menu, organize..............2–61
Map Datum, device................2–56
Map display with data fields..2–65
Map in Track UP....................2–54
Map Information, settings......2–54
Map installation in the GPS 4–110
Map motion.............................2–49
Map reference system.............1–34
Map section, [de]select.........4–116
Map section, send.................4–114
Map spheroid...........................1–34
Map types................................1–12
Map View, BaseCamp..........4–122
Map, send to device..............4–113
Map, slide................................2–44
Map, system setting................2–54
MapInstaller...........................4–113
Maps, Connect......................4–152
MapSource...............................1–20
Marine Chart Mode................2–62
Marine, device set up..............2–62
MARK key...............................2–46
Mass storage..........................4–106
Memory size, micro SD.........1–11
Memory Used..........................2–59
MENU key...............................2–43
Messages..................................2–68
MicroSD card........................4–108
  preprogrammed...................1–12
Mils...2–56
Missing applications...............2–61

Move Point, BaseCamp 4–125

N

Nautical maps 1–18
NAV folder 4–107
Navigation text, map 2–54
NMEA interface 2–53
North reference 1–37
North Reference, device settg 2–57

O

Off Route Recalculation 3–81
OpenStreetMaps 1–16
Operating manual 2–41
Optional routes 3–92
Orientation of the map 2–54

P

PAGE key 2–43
Page scrolling 2–46
Page Sequence, Setup 2–60
Partial installation 4–114
Photo navigation 4–145
Photo tagging 1–33
Playback animation 4–148
Plug-in 4–118
POI (Points of Interest) 1–25
POI folder 4–107
POI Loader 3–95
POIs
   search in BaseCamp 1–26
Position Format 2–56
POWER button 2–42
Power supply 2–41, 2–53
Profile 2–66
Project Waypoint 3–98
Proximity Alarms 3–96
Pulse curve in BaseCamp 4–147

Q

QUIT key 2–42

R

Receive from Device 4–147
Rechargeable batteries 2–41
Record 2–58
Record Method 2–58
Recording Interval 2–58
Remove Hardware 3–90
Remove, BaseCamp 4–126
Reset Trip Data 2–59
Reset, hard 2–48
Reset, Trip Computer 2–65
Reverse Route 3–85
Rework, Track 4–125
Road maps 1–12
Rocker key 2–46
Route navigation 3–80
Route Planner 3–84
Route, convert 4–126
Route, create 3–84
routes 1–22
Routing, set up 3–81
RTCM interface 2–53

S

Satellite information 2–52
Satellites 1–34
Save Portion 3–101
Save Track 3–100
SD card 4–108
Search near 3–83
Serial number 2–41
Setup, device 2–52
Shaded Relief map, turn off ... 2–55
Share Wirelessly 3–91
Show On Map 3–86

Device set up, track 3–93
Sight 'N Go 3–97
Software
    BaseCamp 9
    Connect 9
    Garmin Express 9
    Geosetter 4–142
    POI Loader 3–95
Spanner interface 2–53
Spell Search 2–44
Stop Navigation 3–84
Stopwatch 2–63
System set up 2–52
System structure 4–106

T

TCX file 4–105
Temperature sensor 1–11
Text Size, device setting 2–55
Time format 2–52
Tirol hiking map 1–14
Tone, routing 3–79
Tones, button 2–53
Total Ascent 2–64
Total elevation, barometric .. 4–137
Tour portal 4–118
TracBack 3–99
track 1–23
Track file 4–129
Track from Selected Route, create
    4–126
Track in Device
    zero out track memory 2–59
Track in the Device
    clear current track 3–101
Track Log 2–58
Track Manager 3–86, 3–93
Track navigation 3–86

Track points 1–24
Track properties 3–93
Track recording on the PC .. 4–145
Track, convert 4–126
Track, device setup 2–58
Track, rework 4–125
Track, save 3–100
Track, show 3–86
Track, storage space 3–90
Track, turn on visibility 3–92
Tracks, organize in BC 4–129
Trip Computer 2–63
Trip Data, set to zero 2–59
True 1–38
Turn list 2–60

U

Units 2–52
Unlocking process DVD 4–111
Update device software 1–39
Updates 1–39
Upload from BaseCamp 4–128

V

Vertical Speed 2–64
via points 1–26
VIRB 1–11

W

WAAS 1–38
Wander-Atlas Tirol 1–14
Waypoint icon, create 4–141
Waypoint, delete 2–45
Waypoint, edit 2–45
Waypoint, save 2–44
Waypoints 1–24
Waypoints, delete all 2–59
Windows Explorer 3–89

www.bike-gps.de
    transalp planning tool 4–136
www.earth.google.com
    Google Earth 4–120
www.garmin.com
    Software Downloads 4–111
www.geosetter.de
    georeference photos 4–142
www.gpsies.com
    Tour portal 4–118
www.gps-tour.info
    Tour portal 4–119
www.opencaching.com
    geocache portal 1–32
www.openstreetmap.org
    free map poral 1–16

www.red-bike.de/gps
    Red Bike Tour Portal 4–117
www-addresses
    http
        //connect.garmin.com
            Online evaluation 4–150

Y

yelp 4–132

Z

ZIP folder, unpack 4–118
Zoom keys 2–46
Zoom Levels 2–55

You can find further help in the FAQ (Frequently Asked Questions) section of the Garmin Web site at www.garmin.com > Support > View All FAQs, or in the Garmin Forum at forums.garmin.com.

Overview of all Edition GPS Praxis Books by Red Bike

German versions:

GPS Praxisbuch Garmin Edge705 / 605,　　ISBN 978-1-4461-8831-6;
GPS Praxisbuch Garmin Dakota/ Oregon V2,　ISBN 978-3-8391-7017-5;
GPS Praxisbuch Garmin GPSMap62 – Serie,　ISBN 978-3-8423-2770-2;
GPS Praxisbuch Garmin GPSMAP64 – Serie,　ISBN 978-3-7322-8520-4;
GPS Praxisbuch Garmin Edge800,　　ISBN 978-3-8391-8210-9;
GPS Praxisbuch Garmin Edge 810,　　ISBN 978-3-7322-3028-0;
GPS Praxisbuch Garmin Edge 820,　　ISBN 978-3-7412-8570-7;
GPS Praxisbuch Garmin Montana – Serie,　ISBN 978-3-8423-6706-7;
GPS Praxisbuch Garmin Monterra,　　ISBN 978-3-7322-4589-5;
GPS Praxisbuch Garmin eTrex 10, 20, 30 ff.,　ISBN 978-3-8423-6707-4;
GPS Praxisbuch Garmin eTrex Touch,　ISBN 978-3-7386-2149-5;
GPS Praxisbuch Garmin fēnix3/Chron./epix　ISBN 978-3-7386-2430-4;
GPS Praxisbuch - Tourenplanung mit Garmin BaseCamp,
　　　　　　　　　　　　　　　　ISBN 978-3-8482-2144-8;
GPS Praxisbuch Garmin Oregon 6xx-Serie,　ISBN 978-3-7322-3031-0;
GPS Praxisbuch Garmin Oregon 7xx-Serie,　ISBN 978-3-7412-8555-4;
GPS Praxisbuch Garmin Edge Touring/ Touring Plus,
　　　　　　　　　　　　　　　　ISBN 978-3-7322-8500-6;
GPS Praxisbuch Garmin Edge 1000/Explore,　ISBN 978-3-7357-2486-1;

English versions:

GPS Praxis Book Garmin GPSMAP 64 Series, ISBN 978-3-7386-1494-7;
GPS Praxis Book Garmin Oregon 6xx Series,　ISBN 978-3-7386-5323-6;

www.ingramcontent.com/pod-product-compliance
Lightning Source LLC
Chambersburg PA
CBHW070237230526
45470CB00002B/445